MAKING THE CONNECTION

HOW TO PICK OUT, PICK UP, HOOK UP AND ENJOY YOUR STEREO SYSTEM

By Davidson Corry

Little Free Library!
Always Free!
Never For Sale!

PIPELINE BOOKS   SEATTLE

Copyright © 1975 by Davidson Corry
All rights reserved.
First Printing August 1975
Cover art by Mr. Gary Eagle, Chesaw, WA

Published by Pipeline Books, Box 3711
Seattle, WA 98124

	Introduction 1
1.	A little history, Maestro, if you please 3
2.	The print medium 7
3.	You'd better shop around 13
4.	Choosing a system 19
5.	The ratings game 26
6.	The question of quad 32
7.	Test-driving a stereo 42
8.	Feel the power 50
9.	They float through the air with the greatest of ease 63
10.	Miles of music 72
11.	The loudest ain't necessarily the best 84
12.	Old wives' tales 92
13.	Care and feeding 100
	Appendix 113
	Glossary 119

INTRODUCTION

WHY (YOU MAY ASK), with everyone and his brother coming out with guides to buying hi-fi equipment, why should I buy this one?

Fair question... but this isn't just like the other books. Most hi-fi guides that I have read seem to treat the reader like a customer: Glowing descriptions of systems and equipment, but not much help once you're sold and out the door.

That seems to me to go less than halfway. Making the connection involves not only knowing *what* to buy, but also *why*; where the best (not necessarily the cheapest) place is to buy it; and what to do with it once you've got it home and out of the carton.

Writers, I discover, have nothing original to say. What answers I have compiled to those problems we just mentioned come from friends and teachers I have met in school, at various radio and recording studios around the country... a list too long to print here. All I've added was bad grammar and sloppy syntax.

If I'm going to thank anyone, then, it would have to be the lady that did the dishes so that I could get back to my typewriter. In a way I'm going to be sorry that the book has been published. Now I'll have to start doing the dishes again.

Chapter One

A LITTLE HISTORY, MAESTRO, IF YOU PLEASE.

THE FIRST "HIGH FIDELITY" EQUIPMENT was sold away back in the 1920's—or so the ads for radios with "high fidelity reproduction of the transmitted wave" would have you believe. Surprisingly, though, high fidelity did not grow up with the development of broadcasting, but rather is the descendant of the movie industry.

Radios accustomed people to having music in their homes; but most radios were far from "hi-fi"—and they didn't have to be. It was enough that you could hear Paul Whiteman and the Dorsey Brothers hundreds of miles from where they were really playing.

Talking movies, on the other hand, needed all the "fidelity" and realism they could get, to prove themselves to a skeptical public. Tinny voices and insufficient volume were too likely to snap the audience's illusion of reality. But transistors hadn't yet been invented, and even tube amplifiers were limited to five or ten watts. So the engineers began working, improving the amplifiers, getting more power and better frequency response from them, making speakers more efficient and less tinny. They invented the crossover, the infinite baffle (closed box) speaker and the bass reflex (ported box) speaker. They developed tone controls and equalisation curves and most of the principles hi-fi uses today. But the equipment was too costly for home purchase, and besides you'd need a theater hall to hold the speaker cabinets, anyway.

It wasn't until the post-World-War II technological boom that audio equipment became both good enough and cheap enough for home use. Even so, high fidelity equipment was the eccentric plaything of a few rich fanatics until a way was found to shrink the speaker cabinets.

Speaker cabinets, up to that point, had to be three or four feet on a side to get adequate bass; a smaller cabinet would make the speaker sound tinny and high-pitched. An acoustics professor at New York University took a look at the physics of the system and discovered he could get good bass in a small box just by changing the design of the speaker slightly. Edgar Villchur's "acoustic suspension" speaker caught on, and high fidelity took off.

Meanwhile, the early 1950's brought change to radio as well. Network variety and comedy programs had jumped to the new medium, television, leaving radio stations with a lot of empty time. They began filling it by playing records, which up to that time had been a rarity. Announcers became disc jockeys, and the Great American Top-Forty was born. All this free advertising in turn boosted record sales, and more and more music of better quality became available for playing at home. As competition got fiercer, records and equipment got better and better, seeking an edge in the marketplace.

At first, every record company had its own system for equalising records, and preamplifiers might have four or six different settings for the phonograph input, to get the proper tonal balance. Complaints from hi-fi owners and manufacturers soon pressured the record companies into forming the Recording Industries Association of America, and standardizing recording technique.

Similarly, when manufacturers stopped pooh-poohing the laboratory demonstrations of stereo and realized that it meant selling twice as much equipment, they all came out with competing stereo recording and playback systems, most of which didn't sell or died for lack of money, a few of which are the standards today. Quadrophonic systems are going through much the same game today.

One other contest emerged with both contestants winners. The old 78.26 rpm record standard made for short records and burned out needles too quickly. Two companies came out with records that turned more slowly to get more playing time. Columbia introduced the "LP"—a ten-inch record that turned at 33 rpm and got about fifteen minutes of music on a side. They shortly expanded the size to twelve inches, which gave about 25 minutes per side, and allowed an entire symphony to be put on a single record.

RCA went the other way, putting out a seven-inch **record that** turned at 45 rpm, and held a single song on each side. **This new** single-play record had a large hole in the center instead of **the** usual small one. RCA's plan was to market a cheap, **portable player** designed for the new large-hole record and cash in on **the** youth market. The plan both succeeded and failed beyond **their** wildest dreams.

The RCA players were cheap, but not as cheap as a packet of little plastic adapters—RCA didn't sell too many of the little players. But the convenience of the smaller records sold the public, and the lower costs of recording a single sold the music industry. It became standard practice for a record company to release a single, and on the basis of its success later release (or not release) an album. It was well into the sixties before an album could stand on its own merits.

High fidelity became socially acceptable in the late 1950's; by the mid-sixties some sort of sound system was a social necessity in a properly modern American home. And the post-war boom babies, now grown to college age and affluence, took stereos with them as they moved out on their own.

Rock music grew into an incredible empire—billions **of** dollars, well over half of the entire music industry—and the need **for** stereo equipment grew with it. Stereo had won its battles **in the** early sixties and monophonic equipment was no longer even **sold** generally. The sudden boom in sound equipment brought on with The Beatles and their successors entrenched it so firmly that, when quad made its bid for the spotlight five years later, stereo would not be replaced.

Much of this equipment was cheap and poorly **designed, sold** on features rather than performance—or even simply **on** cost—but audio equipment grew steadily better under **pressure.** Modern technology was working daily miracles, and youth had **no** patience with imperfection.

The equipment has gotten *so* good, in fact, **that professional** studios have been forced to improve their **own** standards just to stay ahead. Quality stereo equipment today **is** almost as good as recording studio equipment, if built a little **less** durably; it's as good or better than most broadcast **station** equipment, and audibly superior to the best of studio **equipment** of ten years ago.

If, that is, you buy wisely and use it properly. **That's why we're** here.

Chapter Two

THE PRINT MEDIUM.

If you're in the market for a hi-fi the first thing you'll probably do is dash out to the newsstand and buy up all the hi-fi magazines in sight.
A word of warning: Don't.
It might seem that the stereo magazines would be the first place to go for information on stereo components, but, with a couple of notable exceptions, they aren't, for a number of reasons.
First, go borrow a copy from the library or a friend—old issues are easy to find. Look through it and see what the articles are all about. I have a typical issue of Stereo Review at my elbow as I write. It contains 11 articles on music, ranging from rock to classical, four book reviews (all about music), two regular columns (about music), one editorial (about music), three sets of record reviews, three columns about various technical subjects, a new-products section essentially reprinting the manufacturers' handouts about their new components, Julian Hirsch's test reports on five components, and *one*—count 'em, *one*—article of general interest to someone buying a new stereo (on how much power do you need for realistic rock reproduction).
Do you see the point? Stereo Review—and most of the other slick-page national hi-fi magazines—are written not for the guy who's about to buy a stereo, but for the guy who already *has* one. It is probably not worth the better part of a buck for the ten or 12 pages that might do you some good in an issue—especially not when you can get a copy out of the library, anyway. You'll probably have to dig back four or five months for the review of the model you were interested in anyway, so why not?

Reason two: If everybody stopped buying the slicks tomorrow, they would be hurting but they would probably not go out of business. The big magazines do not make their money off subscriptions and newsstand sales. The big money is in advertising. The same issue of Stereo Review has 58 pages with advertising on them—not counting the classified section and the "reader service" card (a real neat way for the reader to write away for a bunch of manufacturers' brochures)—out of 120 pages and covers.

This advertising causes some weird effects. For one thing, the equipment test reports are seldom bad. If a component is good, the reviewer usually says so, quite vehemently. If a component is just pretty good, the reviewer usually raves about it anyway, especially if the manufacturer is putting a lot of advertising in his magazine. Good points are pointed out, failings are usually quietly left unmentioned or glossed over quickly. A piece of equipment has to be a real loser before it gets a bad review in most of the magazines—and even then, the reviewer usually tries hard to say *something* nice about it. You never know—they might want to buy some advertising some day.

This is not to say that the reviewers are dishonest. Most of them consider themselves quite honest men, and indeed their technical reporting is usually quite accurate. You can ferret that information out of the figures and charts they supply—*if* you know how to read figures and charts.

But for readers who *don't* want charts and figures, they also have to supply subjective evaluations ("...good clean highs," "crisp transient responses," "plenty of low end") that follow Lewis Carroll's dictum in *Through the Looking-Glass* ("...when I use a word, it means just exactly what I say it means, neither more nor less.") A columnist who gets too harsh and critical in his subjective evaluations soon finds himself shown the door. Otherwise, the publisher will fire both the columnist *and* the editor.

Technical reviews are bullets of hard-hitting, factual reporting, though, compared to the advertisements themselves. These little gems are sheer masterpieces of misdirection, connotation, innuendo and subtle snob appeal. They have to be; when you've got a multi-million dollar business, and you're competing with a hundred other guys with *their* multi-million dollar businesses for John Q. Freak's stereo dollar, you'll use every trick in the book, and a few you haven't written down yet.

The manufacturers hire designers to make their

components look sleek and sexy—complicated enough to look expensive, simple enough to operate easily. They hire advertising agencies to rewrite the technical manuals and replace the important words with the important-sounding ones. They photograph the equipment to make it look huge and towering. They compile and list specifications to make their brand more impressive than the competitors'—and when they run out of specifications, they invent a few of their own. They invent technical problems that their design has "solved," and talk loud and long enough about them that the engineers and reviewers start taking them seriously. And most especially, they use words like a magician uses his wand; to attract, to distract, to amplify, to obscure, to hide, to point.

Some salient examples: Some of the catchy words in speaker advertisements these days are "professional," "studio" and "monitor." These words are used to give the impression that the speaker is used by recording engineers to monitor and evaluate their work. A couple of companies have gone so far as to take the grill cloth off their speakers and paint them grey or flat black, then sell them as "studio monitor" speakers—because *real* professional studios often leave the grill cloths off, and often have the speaker boxes painted grey or black.

Now, paint and grill cloth isn't going to appreciably affect the sound of a speaker—it'll sound the same if you paint the box neon pink with metalflake blue pinstriping. But it sells speakers.

What is a "professional quality speaker?" Is it a speaker that professional recordists might use in a studio? *Could* be. The phrase *could* just as well mean it was a speaker made by people who make speakers for a living—and they could make *bad* speakers for a living. A lot of people do. But the phrase probably doesn't mean a damn thing—it just *sounds* good, and sells a lot of speakers.

Fads are another advertising gimmick. Sometimes the advertisers try to start the fad themselves (like quad). Sometimes the fad gets rolling by itself and the manufacturers have to run like hell to catch up—Dolby cassette decks are a good example.

In the late '60's a Briton named Ray Dolby developed a circuit to reduce hiss in tape recording. Dolby's noise reduction unit gained immediate acceptance in recording studios, but by the early seventies better circuits had been developed, and Dolby's professional market was starting to dry up.

Dolby saw a huge market, however, in cassettes. Despite their convenience, cassette decks were not selling rapidly because of the problem of tape hiss—just what Dolby's black box improved. So our boy Ray developed an inexpensive version of his circuit and licensed it to any manufacturer who cared to pay for it. Within a few years nearly every cassette deck maker was using Dolby in his decks—the few that weren't had either invented similar systems of their own...or they weren't selling many decks.

The point of this chapter is that advertisements, and ad-supported magazines, are fun to read, fun to talk about, but they aren't a totally reliable way to help you pick out a stereo.

I did *not* say "Don't buy the hi-fi magazines." They are informative, useful interesting, occasionally hilarious. There's nothing funnier than "watching" two partisans of some particular classical recordings battling it out in the letters page over which version is best. You will pick up a good deal of information on stereo and music—and a background "feel" for the jargon that will make it easier to understand what stereo is all about. But the slicks aren't much use in getting *started* in hi-fi. There *are* a few magazines that do honest, critical reporting on equipment performance—that will say a unit performs poorly if so it does. *The Stereophile* is the oldest of these. *The Audio Amateur* is another that gets into technical questions, although *AA* tends toward design-and-build-your-own articles. A few of the slicks, such as *Audio*, tend towards good technical reporting and non-music articles—but don't forget the library is a better source than the newsstand.

Making the Connection

Chapter Three

YOU'D BETTER SHOP AROUND.

STEREO IS A HUGE BUSINESS—literally billions of dollars a year. It is also highly competitive, and that can work either for or against you. The people who sell stereos are good merchants—or they don't stay in business very long. They may also be hi-fi experts, but it's not necessarily the case.

To get an edge on the competition, each different kind of hi-fi dealer has a different gimmick, an advantage he can offer his customer. There's also usually some drawback that goes with the gimmick. Knowing how to balance these off is the trick to getting the best buy—not necessarily the cheapest buy—in a stereo.

For example: The little Ma and Pa store down the street sells stereo for higher prices than the SuperStereoMarket in the shopping center—but maybe he's got a better service department than the big store. You have to decide whether you're looking for better service or lower prices.

You don't even have to see the guy you're buying from—there are mail order houses now that sell across country at close to wholesale prices . . . but that's a long way to ship your stereo for repairs.

Let's look at some of the different kinds of stereo stores.

The small specialty retailer

Back when hi-fi first got rolling, a few of the real enthusiasts set up shop in the hi-fi business—it made it easier to get stereos themselves, and it sure beat working for a living.

Quite a few of these are still around today, and there are a lot more younger but similar stores around as well.

Prices in these small specialty shops are generally not the lowest in town, but there are compensations. The smaller stores' salespersons are generally better informed about the workings of the equipment, and can better answer your questions than salespeople in larger stores.

Another advantage of the smaller stores becomes evident when you begin to trade up to better or newer equipment. Almost every kind of stereo store gives a discount on purchase of a whole hi-fi system; the smaller stores are usually more willing to give you discounts on individual components as well.

There is a great American myth that a business must grow or die. Not a few little stores have done neither, but just hummed quietly along at the same pace for years. Some others have grown and branched out to two or three little showrooms. Still others have just gotten bigger, added more lines and more showroom until they've come to resemble

The stereo supermarket

Walk into the glass-and-chrome showroom and you'll see row upon row of stereo equipment—a dozen different lines, accessories and extras scattered around, and three grinning salesmen waiting to pounce on you as you walk in the door.

At this level of merchandising, things start to get a little impersonal. These stores use their size, and the volume of equipment that passes through their hands, to get lower bulk prices on equipment—savings which can then be passed on to the customer. But at the lower prices, profit margin is slim, and every sales trick in the book gets pulled out.

Their main business is in the selling of systems. They prefer to sell an $800 system for $725 than individual components for $800—because the extra time selling three components could be put to better use selling three systems. Therefore you'll usually find individual components sold at little or no discount from list price, giving a better profit margin to make up for the extra sales time.

Even food market and department store tricks are used. In supermarkets you'll see cottage cheese sold at half-price, under cost—to get you into the store where you'll buy other things. It's called a "loss-leader." Same thing in stereo—only it's the phono cartridge that's drastically reduced. Don't forget that "loss-leader" cartridge—we'll be talking about it again in a page or three.

If the store is big enough, it may have its own equipment made for it. "House brand" equipment can be electronic, but the most common house brand component is a speaker. There are a number of factories around the country that design and build speakers for stereo chains. They will build to any specs—really good quality, or as cheap as the traffic will bear. Sorry to say, "cheap" is usually the option chosen.

These speakers will then be given an inflated list price by the factory, so that the store can claim a big discount on a system including the speakers. The list price on the house brand "Gonzo VII" speakers is $250 a pair. The store will generously include them in a $500 system at $150 bucks off. Great—you're only paying $100 for them . . . but the store only paid $40 for 'em, and the factory's doing fine because they only put $20 into them in the first place. Everybody thinks he's made out like a bandit, and you've got some overpriced speakers.

Let's go back to that loss-leader cartridge for a second. Do you think it's a *real* "loss-leader?" Nope. Manufacturer's list price is $49.95, sure, but you're not getting such a deal when you buy it for $20—because the owner bought a hundred of them for $12 each.

Your service facility at a bigger store may be a bit skimpy as well. Service technicians get paid very well, and that's an overhead item the manager can do without. A few big stores do have good service departments, and a reputation to go with them for good service after sale. Most depend on the factory-authorized repair stations in the region—which could mean a long trip to town if your new stereo goes up in smoke. Better to ask first.

This is not to condemn the larger stores—you can get very good deals on equipment if you shop carefully, take advantage of sales techniques and know what you're looking for. The bigger stores are much more in the business of selling than in the business of hi-fi. If you just meet them on their own terms you'll do fine.

Other stores

Radio store chains like Lafayette and Radio Shack sell stereo equipment; they may have a brand of their own manufactured for them in Japan, Korea or Hong Kong. Watt for watt this house-brand equipment is less expensive but also less well-made than the major hi-fi brands. It's one cheap way to go if you just can't afford the good stuff, but steer clear of their turntables—they are usually cheap and badly made.

Some of these stores also sell national-brand equipment as well, especially the stores which are not owned by the parent company but franchised. If the equipment has a good reputation it'll be good no matter where you buy it; but don't forget that the store may not be so helpful when you bring the set back for service. Always get warranty information ahead of time.

Discount stores usually have a stereo section, which is piled high with phono-eight-track AM-FM quad compact combos with little detachable speakers. Avoid them like the plague they are. They will eat your records, scrape the music off your tapes, and give you a tin ear doing it. It plays music, *si*, but not high fidelity music. Besides, if you can only buy it at one place, there's no place to go for a better deal, and you're in real trouble if they ever close up shop.

Mail order houses

The hottest things in the stereo biz. Rent a warehouse, hire a few kids to boot the forklifts around, buy a few ads and you're in business. No fancy showroom, no demo models, no breakage, no salesmen, no messy overhead to deal with. Result: Low prices for the customer.

Also result: Nobody to turn to when it breaks. Many regional service centers will not repair equipment, even if they have the service agreement for that area, even if the warranty card is filled out properly, when it was bought by mail order. And hassling the problem out by mail between Elkhart and Cucamonga means a month or more of no music while you agree on who's going to fix the little sucker.

Again, you're paying for the dance, you call the tune. Mail order prices are lower than just about any other stereo outlet's prices; if the mail-order house has a good reputation, your chances of trouble are small. Kit manufacturers have been selling by mail-order for years with great success and little customer complaint. But if you do run into trouble, don't say we didn't warn you.

Sad to say, stereo equipment is expensive. Where does all the money go? To a chain of people who sell the equipment to each other before they sell it to you.

The manufacturer sells his product to a wholesaler, a man with a lot of warehouse space and a lot of trucks. He then takes a profit and sends it on to the regional distributor.

The retail dealer buys from the regional distributor, puts it in his showroom, and you walk out with it under your arm one night, $500 lighter in the bank account, having paid for braces on the teeth of four families.

There are other steps in the chain, too—prices may vary if the equipment is bought for home, professional, or educational use. But prices stay pretty stable—and profits substantial—down to the regional level. At the retail level it's something else agin. Price wars and competing "Special sales" are commonplace. Cutthroat competition in stereo sales (and other fields) threatened to put a lot of retailers out of business before the introduction of so-called "Fair Trade" laws.

Equipment that is fair-traded has a list price set for retail sale by the manufacturer, and in states with fair trade laws the equipment may not be sold for under that price. The intent was to give the smaller merchant a fair chance to sell his goods against the bulk buyer who could afford to undercut him. Unfortunately, a lot of brands have used fair-trade to insure a good and healthy profit for their wholesalers and retailers (and themselves, natch).

Most major brands of equipment have fair trade pricing, component equipment especially. A few of the major speaker brands are fair-traded, but most are not. Similarly, phono cartridges are usually not fair-traded; turntables may or may not be; accessories and adaptors seldom. Which makes for a convenient dodge on the part of the retailer. By selling a system to you, he can quote full price on the receiver and take a correspondingly large discount from the speaker price. In effect, you're getting a price break on the fair-traded item; but on paper, it's legal.

It may seem a bit bass-ackwards to tell you *where* to buy before you know *what* to buy. Fear not—now you know what to look *out* for, and can spend your time thinking about what to *look* for.

Chapter Four

CHOOSING A SYSTEM.

THERE ARE SOMETHING OVER A HUNDRED DIFFERENT major brands of turntables, tuners, amplifiers and receivers on the market today. With such a bewildering array of equipment to choose from, where does the hi-fi buyer start?

The first step is easy—look at your wallet. Figure out how much money you can afford to spend on a stereo system, and *put it in the bank*. This is a trick I've learned through hard experience—you're going to be looking around at stereo equipment for a couple of weeks, minimum, and too many people get bored and disgusted with the search, ending up spending the money on pizza and beer. By the time they've decided to buy a certain system, they can't afford it; and by the time they've saved enough so they can afford it, they're not sure they want it after all.

How much can you afford? More than you might think. Sure, $500 is a lot of money to plunk down on one go—a thousand even worse. But stop and think a minute—chances are you'll keep your music system for two or three years, or trade it up to a better system and get half the purchase price back on it. A thousand dollars spread out over three years works out to about a buck a day—you're probably spending more money on records every week already.

Consider also that you're investing in good sound for now *and* the future. Buy a decent system now and your record collection will be in good shape when you get your Dream System. Buy a schlock system now and your records will be chewed up and ruined—even a good system won't get good music from them later, because the cheap stylus and tone arm will have scraped the music right out of the grooves.

It varies with experience and circumstance, but a good figure to shoot for on your first system is about $400 or $500. Less than $400 means you'll be skimping on some aspect of the equipment, getting poor sound and possibly damaging your records.

More than $500 is probably too much for your first system, too. It sounds absurd, but most laymen *can't tell the difference* between a $500 system and a $1000 system . . . because they both sound so much better than the tinny little car radios they're used to. By the time you've lived six months or so with your system you'll be able to pick out the little deficiencies that make the $1000 system worth twice the price—and by that time you can afford to trade up if you like.

There's another reason—the improvement in quality doesn't go up as fast as the price. A $500 system sounds a hell of a lot better than a $250 system; a $1000 system sounds noticeably better than that. For $2000 you can buy a system that is just a little better—beyond $2000 you have to have the classic Golden Ear to tell any difference. Each step towards theoretical perfection gets harder and harder to take, and the price goes up accordingly. There's no sense in paying for sound quality that you can't yet even hear.

Okay, you've settled on a budget for your system, and it's time to go compare brands. Here you're helped by the fact that everyone makes pretty much the same sort of equipment. Any two receivers within fifty bucks of each other probably have the same power and performance specs, or so nearly identical that the difference is inaudible. Same deal for turntables, decks or other components. Your next step is to decide what fraction of the budget to give to each component.

The first component to choose is a turntable. At this point the neophyte hi-fi enthusiast usually balks at spending enough money, fearing he'll skimp too much on amp and speakers. It seems more logical to divide the money up so that you can get the most power for your money.

Remember that a cheap turntable can destroy your record collection, even with two or three plays. Your records are going to be with you for a long time, and it's worth while protecting them. My advice is to spend not less than $100 **on the turntable, and** more if you can possibly do it. Two **hundred dollars** is not out of line for an automatic changer; manual turntables in the $100 range are also reliable. Seventy-five dollar wonders will very quickly begin to show up harshness and distortion in the records, and once you've heard it, it's in the grooves—the only cure is to replace the record.

That leaves us maybe $350 to split between speakers and receiver. If you can get along without FM for a while, it may be a good idea to get an integrated amp (or preamp and power amp) alone, and buy a tuner later. This is a trifle more expensive in the long run, but lets you get good sound right from the beginning.

Speakers are going to be the hardest—and most frustrating—articles of your search. Even national brands priced at the same level vary from good values to complete ripoffs—and there's no way to tell from just looking at them. Critically listening to a speaker's sound, such as we'll discuss in the next chapter, is essential to getting a good value. In general, the smaller speakers will be more *efficient*—that is, you'll get more volume from a given amplifier than larger speakers. What you probably *won't* get is really low bass, even if you turn the bass control all the way up on your amp—the speaker elements are too small to reproduce it properly. This is less important than it sounds, because most records produced today don't have all that much really low bass—they're recorded knowing that most people don't have really premium speakers, and the frequency response is tailored accordingly. If you're really a gut-shaking, brain-homogenizing bass freak, be prepared to shell out for it. Bass requires both higher power and bigger speakers... and that, my friend, will cost ya.

In general, find an amplifier in the fifteen to thirty watts RMS per channel range, then find a pair of speakers which sound good and which will give you enough volume at the power level of your receiver or amp. You may find a pair of speakers that are very efficient, and you can get by with lower power and price on the amp. You may have to buy a relatively powerful amp to drive the clean-sounding but inefficient speakers you've chosen. Find a good amp-and-speakers combination, then add your chosen turntable and you're in business.

"How much should I spend on my stereo?"

"$500.00."

Things are obviously not as simple as that, but it gives us a good place to start from. How much *you* spend on your system depends on your listening habits.

What sort of music do you like to listen to? Rock music tends to require more power for good reproduction than middle-of-the-road and most of the classical repertoire. Of course, there's no reason why you can't play Joni Mitchell louder than Bloodrock, but most people don't. Some classical pieces require a lot of power, too—for sheer volume the *1812 Overture* and *Night on Bald Mountain* can hold their own with Grand Funk any day.

Power requirements go up sharply during fortissimo passages in orchestral music, though you can get by with less power by just turning the volume down. Rock music is generally recorded at a constant volume level through the song, but the electric and electronic bass notes may require a great deal of power for proper reproduction. If you want to reproduce that bass at realistic levels, you may need fifty or a hundred watts of power per channel—and you'll have to pay for it.

How loud do you listen? If you turn the volume all the way, naturally you'll need more power. Five watts per channel may be completely adequate for the little old lady who listens to Mantovani all day long; twenty watts per channel is fine for most people; forty or fifty watts gives you all the extra margin you need for casual listening, while the synthesizer freak down the street may not be happy with his monster amp, the one that dims every light on the block when he turns it on.

How do you set the tone controls? Turning up the treble adds negligible power drain, but the bass control can make or break your system. Most of the acoustic power contained in a sound is in the bass region—turning up the bass knob all the way may *double* the power drain. The music will *sound* louder—but chances are it will also sound muddy, distorted, and the volume may even fluctuate with the rhythm of the bass notes. If that happens, you need a bigger amplifier.

You may think that you can get along with a little muddiness in the music. Maybe you can, but your amp can't. You're pouring close to the limit of power through the output circuit, and you may be overheating your amp in doing so. You'll have less trouble and better sound if you go to the bigger amp in the first place.

Can't afford a bigger amp? Then find more efficient speakers. In doing so you may have to give up some of the really low bass—don't turn up the bass control to compensate, or you'll be right back where you started.

You may find, in fact, that listening to music at lower volumes and with less bass is real pleasant. You'll find yourself responding less to Da Big Beat and more to the tasty licks the guitar player is trading with the steel man, stuff that the bass distortion fuzzed out before. And by the time you can afford a system that plays good *and* loud, you'll really know how to enjoy the music.

Juggling speakers to fit an amp is only one way to fit a system into your budget. Kit amplifiers and tuners go a long way to getting good sound out of few bucks. The major kit makers have spent a lot of time and money making their kits reliable and foolproof, so don't protest that "I don't know anything about building electronic equipment." The hardest part of building a kit is deciding to try; and even if you chicken out, you can usually find a friendly solder freak who'll built the kit for you. You may have to pay him a few bucks to do it, but the total cost should still be less than a manufactured unit.

Kit *speakers* are also available, very much less expensive than the already-built kind. At least one company makes a complete line of kit speakers, from bookshelf models to studio monitor quality designs.

Once you've done some preliminary looking and have decided what power range and price range you're looking at, it's time to compare brands. This is the fun part.

Nearly every store will have a rack piled high with brochures and literature about its equipment. Grab a handful—reading equipment brochures is as much fun as listening, and probably more useful. Certainly cheaper, and the store manager doesn't mind; every time you look at the pamphlet, it's that much more advertising for him. But don't overdo it—some people get a reputation for wandering into stereo stores and grabbing one of everything without ever buying. This is a good way to get yourself thrown out of the store next time you come in. It will get you labeled as one of those guys who can quote obscure specs at the drop of a stylus, but who plays his own albums with a straight pin and a Dixie cup. Not a good rep to have when you're looking for help.

Ask the salesperson for help. Sure, he's after your money.

...**but he** also knows he won't get it until you're **convinced you're** getting the best deal. Let him know that you're seriously looking for a hi-fi, that you want to spend x dollars on it and you want it to do this, this, and this. Then he won't waste time showing you systems you can't afford or systems you don't want.

Most people who come into stereo stores come with little idea of what they're looking for and end up getting dragged from system to system until everyone's confused. Make it clear that you're there to buy, and you know what you're looking for, and the salesman is much less likely to start horsing around. He knows that he'll lose your business if he does.

Going over the spec sheet with the salesman is often a good idea. Not only will you get a better knowledge of how the equipment works, but it's also a tip-off to whether the salesman knows what he's talking about as well. If you're not sure of him, ask him about a specification that you pretend you don't understand. You'll soon find out whether he knows, *thinks* he knows, or knows he *doesn't* but is trying to fool you.

Ah! You say you aren't sure *you* understand specification sheets? Well, then, just step right this way...

Chapter Five

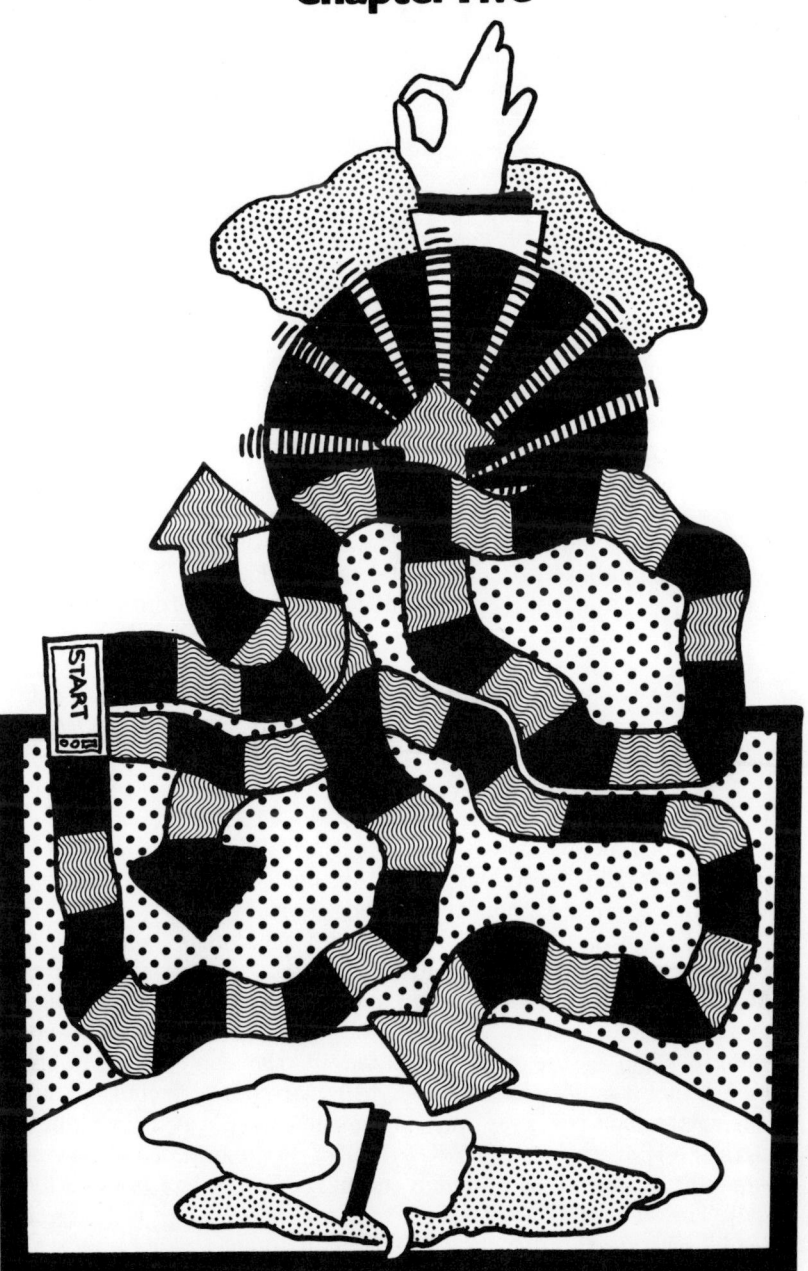

THE RATINGS GAME.

"Figures don't lie, but liars use figures"—an appropriate quote to approach high fidelity specifications. The performance of audio equipment can be measured by how closely it approaches theoretical perfection. Where everyone uses the same standards of comparison, specs can be a useful guide to performance.

Unfortunately, high fidelity marketing is a very competitive game, and some manufacturers use different methods of measurement in an attempt to get better-looking numbers with which to sell their equipment.

The best example of this is in power ratings. The most useful power rating in terms of amplifier performance is the *RMS watts per channel* rating. "RMS" means "root-mean-square," and is a way of determining the effective power generated by a varying voltage through the speaker. The varying voltage is the audio signal being played, and the RMS rating calculates the power your amp is capable of putting out steadily for hours on end, with a steady tone input to the amp.

Unfortunately, RMS ratings are also the most conservative, since they deal with a steady, unvarying tone. Music is not steady; it jumps up and down in volume and pitch, and in between loud passages your amp has a chance to "rest" and radiate some of the heat generated by the components. This means that it is possible to get slightly more power from the amp during the loud passages, and the "music power" rating is higher than the RMS rating by 50 % to 100%. Same amp, same power capabilities, but all of a sudden the paper rating has jumped by half. The letters "IHF" stand for the Institute of High Fidelity, a trade organisation of high-fidelity equipment manufacturers.

26

Being manufacturers, they naturally want to make their equipment appear as powerful as possible, and so a music-power type rating system was adopted by the IHF years ago. By depending on the intermittent nature of music, the IHF rating allows higher numbers than the RMS rating.

The ultimate development of this approach is to calculate the power that could be theoretically developed for a tiny fraction of a second by a sharp pulse fed into the amp. Since the pulse is gone almost as soon as it appears, the circuitry hasn't time to overheat, and the instantaneous peak power (*IPP*) rating can be twice or more the music power rating.

By judicious juggling of these methods you can claim just about anything for your amplifier. Suppose it's got 30 watts RMS per channel. That's maybe 50 watts per channel music power—over 100 watts IPP power. And there's two channels, so we add them and find our 30-watt amp is all of a sudden a "200-watt IPP power amplifier."

Two other factors figure in power ratings. Most amplifiers have a single power supply for both channels (or all four channels in the case of a quad amp). It's possible to get a slightly higher output from one channel by shutting off the other channel(s). Of course, when you're using the system all channels will be operating simultaneously, but for the purpose of inflating powerclaims, let's just run one and then multiply it by the number of channels.

For years, stereo manufacturers used all these rating systems and more to get the best possible power "rating." The conflicting and misleading power claims finally forced the Federal Trade Commission to crack down; the FTC now requires power specifications to be made in RMS watts per channel, all channels driven, at a given distortion level and continuously operating over a relatively long period of time—in other words, how much power the amp is capable of in use at home. However, the manufacturers are not prohibited from using the other methods for rating their equipment, just required to put the FTC rating there too. Look on the back page for the most conservative power rating, listed as "RMS per channel" or "FTC rating" and you'll have a fair comparison with other amplifiers.

A theoretically perfect amplifier would increase the voltage or power of an audio signal without changing the form of the wave even slightly. No practical amplifier is perfect, however, and the measure of its imperfection is in the *distortion* it adds to the wave. Distortion comes in two kinds, *harmonic* distortion and *intermodulation* distortion.

When a single tone is sent through an amplifier, any distortion added shows up as new tones at multiples of the original frequency—*harmonics* of the original signal, which is called the *fundamental*. Distortion is measured by checking first the total power output, then the power output at all but the desired frequency. The percentage of power showing up at the unwanted frequencies is the distortion figure of the amp. The human ear can detect distortion in a single tone of about 1%.

Music and voice aren't single tones, however; they're complex mixtures of different frequencies at different levels, and an imperfect amplifier distorts them differently. In addition to generating unwanted harmonics, the amplifier also generates new signals at the sum and difference of the input frequencies.

Suppose we put signals at 60 Hz. (the frequency of the AC voltage in the wall plug) and 400 Hz into our amp. A perfect amp would give us 60 Hz. and 4000 Hz. in the output and that's all. Our imperfect amp gives us 60 and 4000, but it also gives us 3940 Hz. (4000 minus 60), and 4060 Hz. (4000 plus 60). It probably is also giving us 8000 Hz. (harmonic distortion of the 4000 Hz. signal), 120 Hz. (harmonic of 60 Hz.), 8120 Hz. (8000 + 120), 7880 Hz. (8000 - 120), and assorted other sums and differences of fundamentals and harmonics.

This sort of distortion is called *intermodulation* distortion, and is particularly grating to the ear. Harmonic distortion products show up as octaves, fifths and thirds—nice chord harmonies with the fundamental frequency. Sum-and-difference products, on the other hand, have nothing to do with the harmonies of the music, and sound like harsh, off-key, out of tune signals that really mess up the music.

Intermodulation (or IM) distortion is also measured as the percentage of the total output power that shows up at unwanted frequencies. Distortion figures of .5% to 1% are quite noticeable; .2% is detectable but not particularly annoying. Distortion at or below .1% is pretty well negligible, and well designed modern circuitry is capable of better than .005% distortion.

A note about speakers: Speakers introduce their own distortion, usually 2% or more; but the mechanical distortion of the speaker sounds different than the electronic distortion in the amp, and does not mask it.

Another practical problem with real amplifiers is *noise*. At any temperature above absolute zero, the electrical charges associated with atoms rattling around in electronic components generates *white noise*—the technical term for electrical energy randomly scattered through all frequencies and levels. This is the random noise you'll hear between stations on the FM dial; random magnetic fluctuations cause the same sort of hiss to appear on tape recordings.

If the noise level is sufficiently lower than the signal, it will be *masked*, and you'll never hear it. The power level of the noise is measured in *decibels* below the signal power level. The decibel is the unit of measurement for relative power levels, and corresponds roughly to the smallest change of level that the human ear can detect (a fuller discussion is given in the Appendix).

A noise reading 60 decibels below the level of the music is generally considered inaudible. Remember, however, the noise specification is usually taken referenced to maximum power output for the amp—in most cases you'll be running the volume five to ten decibels below that level. Therefore, look for a noise specification of -65 db. at least, and -70 or better for real quality.

Quite often specs for hum are included in the noise figure. The power in the wall socket is an AC voltage of 117 volts RMS at 60 Hz. On an oscilloscope, it looks like a sine wave that alternates between peaks of about 160 volts positive and negative.

The power supply of your amplifier turns this alternating current into direct current, filtering out the 60-Hz. tone as it does so. How well this hum tone is filtered determines the hum level in your amp. The better the filtering, the less hum.

Hum may show up as a low bass tone, or as a raspy buzz, or both together. The buzz is caused by harmonic distortion of the hum, putting harmonics at 120 Hz., 180 Hz., and every 60 Hz. up to 20,000. Noise, on the other hand, is not steady—it's a constantly-changing rushing sound, like the noise a waterfall makes. Both noises and hum together should be -65 db. or better.

Frequency response tells how evenly the amp amplifies each separate frequency.

Frequency response is measured in bandwidth over a given decibel range. A rating of "±1 db. 20,-20,000 Hz." means that amp will give the same gain at any frequency in the audible range within a decibel of the mid-frequency gain. A decibel is the

29

smallest amount of change in level that the human ear can detect: Someone listening to this amplifier would say it amplified every frequency the same—that it was "flat."

The range of human hearing is from approximately 20 Hertz to 20,000 Hz. Twenty Hertz is a note just below the lowest key on a piano; the highest-pitched musical instruments hit 4000 Hz. or so, with harmonics and transients out to 20,000 and beyond. These harmonics and transients are the sharp clicks and whines that define the edge of a cymbal crash or the sound of a violin.

Modern circuitry is capable of much better performance than "±1 db., 20 to 20,000 Hz." Some purists maintain that they can hear the difference at the extreme ends of the audio spectrum, and extend response to 10-60,000 Hz. or beyond. For most of us, that's creative overkill, but no hassle because "±.1 db. 5 to 100,000 Hz." is still going to be equally flat over the 20 to 20 range. "Twenty to twenty" is still a good standard to set for frequency response.

Power bandwidth is another measure of frequency response—a better one than most, because it specifies the frequency range over which full rated power can be counted on, essentially guaranteeing good response at any volume setting. Unfortunately, this test is usually only reported by manufacturers who know they have nothing to worry about in performance, anyway.

These four—noise, distortion, power, and frequency response—are the major specifications to look for with amplifiers. Noise should be -65 db. or better (the higher the absolute number the lower the noise). Distortion should be less than 1% and preferably better than .5%. Frequency response—"twenty to twenty" or wider, with a plus-and-minus range of no more than a decibel. Power is your problem—just make sure you've got enough to get your speakers going.

Chapter Six

THE QUESTION OF QUAD.

They say that those who do not learn the lessons of history are doomed to repeat them. The high fidelity manufacturers learned history's lessons very well—so well that they're doing their very best to repeat history anyway. The switchover from mono to stereo was highly profitable, and that's a nice history to repeat.

The very first stereo records were matrixed, just like the first quad records; as the FCC had to choose between a number of competing proposals for a method of broadcasting FM stereo, so it is now considering different ways of doing quad on FM.

What has changed is the market. Where early stereo makers sold their products to a public that knew only how exciting the idea of quality music in their own homes seemed, today's buyer knows something of what he's getting into—probably because he's got a $500 stereo system that he's not all that anxious to throw out and start over.

There's no denying (though the advertisers try) that quad is more expensive than stereo. You'll need twice as many speakers, twice as many amplifier channels. One tuner and one turntable are sufficient, but the extra expense of a better phono cartridge and decoding circuitry makes up for it.

When talking about quad, the engineering concept of a "trade-off" often comes up. You don't get something for nothing. If you want to tuck four channels of music down into a two-channel record groove, you have to give something up—either separation between the four channels, or frequency response, or bandwidth, or some other characteristic.

The simplest "compromise," of course, is to keep the four

channels separate in the first place. This system is called *discrete quad*. The only truly discrete quad devices on the market are tape decks, which lay four separate bands or *tracks* of signal down side-by-side on magnetic tape. The "compromise" lies in the extra expense of two extra windings in the tape heads, extra circuitry for the third and fourth channels, and poorer noise performance with the thinner tape tracks. (Professional four-channel tape machines use half-inch tape to get around the noise problem.)

The advantage of discrete quad is better separation between channels, which gives more precise positioning of the sound "image." JVC's CD-4 process puts a four-channel signal into a two-channel record groove while maintaining the same effective channel separation. Though it doesn't fit the dictionary definition, no one minds when JVC advertises CD-4 as "discrete quad," since the effect is the same.

CD-4 makes its compromise by increasing the bandwidth of the signal in the groove. A normal signal has a top frequency limit of 20 kilohertz; the CD-4 groove may have frequencies as high as 45 kilohertz. These higher frequencies contain the information needed to separate front and rear channels from each other.

In CD-4 recording, the left front and left rear channels are electronically added together to form a "main left" signal, which is what you hear when playing the record on a normal stereo player. The system is *compatible* because you don't lose any of the left channel information when playing back in stereo.

A "left front minus left rear" signal is also formed, and this signal frequency modulates a *subcarrier* at 35 khz.—too high for the ear to hear, but easily decoded by the CD-4 circuitry in your amplifier. The sum and difference signals are then recombined to recover the original front and rear channel signals for four-channel playback. The process is similar to FM stereo multiplexing, except that there are *two* multiplexed signals in the stereo groove, one for right channels and one for left.

The disadvantage of the system lies in the requirement for 45 khz. frequency response from the phono cartridge—that takes a cartridge with specially-wound coils and an exotic stylus shape. One such stylus shape was developed by an engineering team led by a man named Shibata, for whom the stylus is named; other similar styli are called "Quadrahedral," or other names. Whatever the name, the special styli and cartridges are more expensive than most stereo cartridges.

The engineers at other manufacturers weren't happy with

the technical problems associated with wider bandwidth. Instead, they chose another tradeoff: Channel separation. If you settle for somewhat less channel separation—and the resulting hazier definition of sound position—you can squeeze four channels into two with relatively simple *matrix* circuitry.

There are two main types of matrix quad: The so-called "regular matrix," typified by Sansui's QS system, and the "phase matrix" such as the CBS/Sony SQ system. Each uses a fundamentally different technique for putting the signal in the groove. To understand them, let's look at a basic two-channel stereo groove first.

The normal stereo groove has two walls aligned at right angles to each other, and at a 45° angle to the surface of the record. One wall carries the left channel signal, the other the right channel.

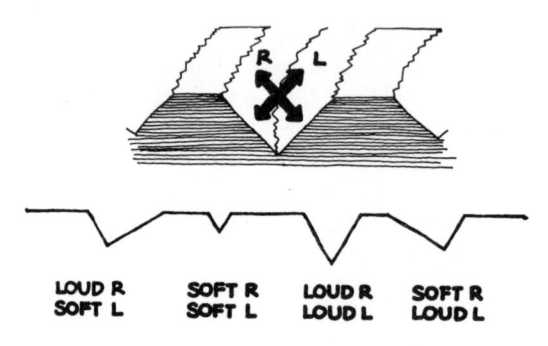

LOUD R   SOFT R   LOUD R   SOFT R
SOFT L   SOFT L   LOUD L   LOUD L

If you didn't fall asleep in high school physics class when the teacher was talking about vectors, you'll remember that two forces acting at right angles to each other don't interact; each wall kicks the stylus in the proper direction without interfering with the motion imparted by the other groove wall, and the two stereo channels go their separate, merry ways. So far so good.

Both phase matrix and regular matrix leave these same motions for their front channels, left and right. The *regular matrix* system codes one *rear* channel as a *vertical* movement of the stylus tip, the other rear channel as a *horizontal* movement.

Here's where the loss of separation comes in. To lift the stylus vertically, *both* walls of the groove have to move "upward" at a 45° angle at the same time; dropping the tip vertically, both move down. For horizontal movement, one wall moves up while the other moves down, depending on the horizontal direction.

(**Students of vector** math may at this point **mumble intelligen**tly **to themselves about** "vector components.")

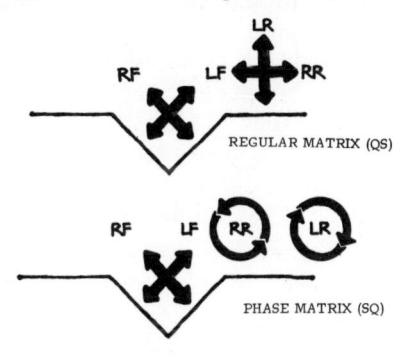

```
          LR
   RF  LF ✢ RR
      ╲╳╱
       ╲╱  REGULAR MATRIX (QS)

   RF  LF ⟳ ⟲
      ╲╳╱
       ╲╱  PHASE MATRIX (SQ)
```

Well, the cartridge doesn't know those are the rear-channel effects—it thinks you've sent two signals to the front speakers, as well as giving a rear-channel signal. So a fraction of the one rear-channel signal shows up in both front speakers—"leakage" from the rear channel to the front, giving low channel separation. The only speaker that doesn't "hear" the signal is the one reproducing the signal at right angles to the vertical tip movement (the horizontal signal, going to the other rear speaker).

The phase matrix system gets a little more tricky—it encodes each rear channel as a *circular* movement of the stylus tip. The left rear channel signal moves the stylus in a clockwise circle; the right rear signal moves the tip in a counter-clockwise circle.

"Well, that," some bright young student is saying, "means that the rear channels are exactly the reverse of each other."

Exactly; and what's worse, the "reverse" or negative of an audio signal sounds exactly like the positive. If you were listening to the rear speakers only, you would hear a monophonic signal with the speakers connected out of phase—there is essentially no separation between the rear speakers at all! The advantage is a slight increase in separation between front and back speakers, since the components of the circular tip motion don't show up as strongly in the front channels as the regular-matrix rear channels do.

The matrix manufacturers claim the advantage of not requiring a special cartridge for playback like the CD-4 records do. This claim has to be taken with a grain of salt. It's true that the matrix records don't increase the signal bandwidth put into the groove—there is a 20 khz. maximum, instead of the 45 khz. maximum in a CD-4 record.

However, the extra gyrations of the stylus impose severe difficulties on keeping the needle in the groove—the circular and vertical groove motions tend to kick the stylus tip right out. It's usually necessary to go to a more expensive cartridge and stylus that will track better at lighter pressures, even though the exotic stylus shapes needed by CD-4 are not required by matrix records.

In listening terms, the QS regular matrix gives fair separation between speakers at adjacent corners of the square, and complete separation between speakers on the diagonals. The SQ phase matrix gives essentially no separation (but an out-of-phase condition) between the two rear speakers, fair separation between front and back and good separation between the two front speakers.

In use, both matrix systems sound pretty good, but neither can hold a candle to true discrete playback. The matrix systems depend a lot on something called "psychoacoustics"—the fact that what you *think* you hear is not always what you hear.

Unfortunately, a lot of "psychoacoustics" depends on advertising—making you think you heard what they *wanted* you to think you thought you heard, even though that's not what you heard or even what you *thought* you heard (huh?).

Example: If you're in the middle of a square of speakers, and a signal is coming strongly from the front left speaker and less strongly from the front right and left rear, your mind will locate the sound source somewhere vaguely towards the left front. If someone asked you where the sound came from, you would probably point to the left front speaker—but if the sound came from that speaker alone and from nowhere else, there would be no doubt in your mind. In an A-B comparison between matrix and discrete quad for sound imaging, the matrix comes in a distinct second.

When you're in the business of selling hi-fi equipment, second place is as good as no place at all; so the marketing boys sent the engineers back to the lab to see what they could do. What they did was to discover another psychoacoustic effect: If you hear two or more sounds at once, your mind only pinpoints the location of the loudest sound. If the engineers could momentarily improve the separation on the loudest sound in the four channels, they could make the overall definition seem improved, even if they did it at the expense of definition on the weaker sounds.

Which is exactly what they did, using amplifiers that sensed the levels of the signals being fed through them, and automatically adjusted the gain of the circuit accordingly. These gain-riding amplifiers were promptly incorporated into the decoder circuitry, dubbed "full-logic" by the marketing boys, and rushed to a breathless public. (The word "logic," by the way, has nothing to do with digital circuitry or computers—it just sounds nice and modern and scientific. They used to use "galvanic" the same way back in the 1800's . . . sold a lot of patent medicine with it, too.)

Most of the quadrophonic records sold today were not specifically designed for four-channel playback—the great majority, in fact, were recorded before the development of quad techniques, and quad releases of older LP's are often four-channel remixes of the original stereo masters. Occasionally a conscientious producer digs out the original 16-track master—if it still exists—and mixes down from that. Newer recordings, of course, can be mixed to quad and stereo as they are made; but even with new material the producer runs into a problem: Compatibility.

A four-channel LP (whether matrix or CD-4) must sound good in quad. Since most of the equipment in existence is still stereo, it must also sound good (and maintain adequate channel separation) in stereo. And, if it's going to get any airplay on the AM Top 40 stations, it has to sound good in mono as well.

This puts severe limitations on how extravagant the producer can get in his four-channel mix. Of course, it's possible to release two (or even three) different mixes, one for each medium—but that's an extra expense that most record companies can't afford on any but the most successful of releases, and certainly not on a new and untried album. Most record producers have not yet gotten the hang of a natural-sounding four-channel mix without the cutesy bounce-the-sound-between-the-speakers tricks, either.

So much for the theory of four-channel sound; the proof of the pudding lies in putting the theory into practice. To make things easier for the present owners of stereo equipment, most of the quad manufacturers have come out with outboard decoders and add-on amplifiers that use the existing components as the front half of the new quad setup.

Outboard matrix decoders almost universally attach to the tape deck outputs and inputs on your present stereo amp—these feed the preamplified signal (but before tone controls or volume adjustment) to the decoder, which splits off the rear channels and sends the front channels back to the tape inputs. Pushing the *tape monitor* switch on your amp connects these new front channel signals to the tone/volume control circuitry and thence to the power section and your speakers. Most of the outboard decoders include a tape monitor switch themselves, so that you don't lose the facility for recording when going to four-channel operation. This monitor is usually two-channel, so that you record the undecoded (matrixed) two-channel "quad" signal.

CD-4 decoders are somewhat different. Since they require the full 45 kilohertz output of the phono cartridge, they are connected between the turntable and the amp, and contain their own preamplifier circuitry for the low-level phono signal. The output of a CD-4 decoder is at high level, like a tape deck, and is connected to the "aux" inputs on your stereo amp, or, if you prefer, to the tape inputs (but you lose tape monitor facilities if you do). Some CD-4 demodulators offer tape outputs for a discrete four-channel recorder as well. Finding two more channels of amplification for the rear speakers can be a problem. A few manufacturers offer "add-on amplifiers" for the rear channels, with volume and tone controls but no input selector switch. Some decoders include rear-channel amplification, usually at a relatively low power-level. To adequately balance a high-powered stereo amp you may have to get a second preamp and power amp for the rear channels to obtain proper balance and tone control.

The most annoying drawback to "adding-on" is its inconvenience in operation. The outboard decoder may (or may not) have a "master" volume control that adjusts the volume of all four channels simultaneously. If it does, you'll still have to pre-set the front and rear volume controls for good balance before using the master knob. Almost certainly you'll have to use two sets of tone controls. And readjusting balance to compensate for program material or different speakers may have you cross-adjusting five or six controls at once.

If you want convenient operation and four-channel sound, there is little alternative to getting a complete four-channel amp or receiver with decoding circuitry built in. Depending on which brand you choose—and who they've paid to use whose devices—you can get four-channel equipment that will properly decode any of the various quad standards, regular matrix, **phase matrix**, or CD-4.

Choosing a four-channel amp or receiver is very similar to choosing a stereo machine, the primary difference being the price. The price of a new quad receiver of good quality is enough to put most people off very quickly, but don't fall into the trap of thinking that an add-on system is less expensive. True, adding-on will involve less cash outlay at the beinning than trading in your stereo system on a complete quad system. However, with an add-on system, after a few years you will have an old, worn-out stereo, and a decoder and rear-channel amp with very little resale value. Trading or selling your stereo *now*—while it's in good condition, and while stereo is still a worthwhile item—will save you money in the long run even though the cash involved may be greater.

One nice thing about quad—you don't need quite as much power per channel for the same amount of volume. For example, if 20 watts RMS per channel was adequate for your listening in stereo, a quad amp with 15 watts RMS per channel will probably do quite nicely. You still need enough power to drive your speakers properly (which is why you can't just halve the power level on each channel); but the volume builds up faster with four speakers than with two. The buildup is most pronounced in the bass, so you probably won't need to boost the tone control or loudness switch as much—another saving in power need.

Experts are still arguing whether you need identical speakers in all four corners, or whether you can get away with less expensive speakers in the rear. For a true quad performance, designed and intended for four-channel playback, there's little doubt; identical speakers all around are needed, otherwise the tonal balance suffers (since each speaker model has its own distinctive "sound").

As we've said before, though, most of the music in the record racks is *not* designed for true-quad reproduction—the rear-channel signals carry "ambience," echoes, spatial effects which don't carry much power in the bass and can therefore be adequately reproduced on smaller, less expensive speakers. Identical speakers are to be preferred, since you're ready for anything, true-quad, near-quad, or just stereo sound-in-the-round; but if you can't afford four $200 speakers, it won't hurt badly to get away with less in back.

Of course, nobody said you had to locate the "rear"

speakers in the rear. Often the quad effect will be enhanced if your "rear" speakers are located to the *side*, or even all four speakers in a semicircle in front of you.

For the severely impecunious (read "broke"), there is a way to get a simulated "quad" effect with two extra speakers and a couple of resistors. Some years ago the president of Dynaco, David Hafler, came up with a way to interconnect four speakers to a stereo amplifier that would matrix-decode a properly-mixed stereo record. Even with records that were not recorded using the Hafler system (as most are not) there is a pleasing full-room sound.

The left and right speakers are connected normally, and placed to left and right and slightly in front of the listener. A *center front* speaker is then connected between the plus terminals of the amplifier, and a *center rear* speaker is connected between the common (ground) side of the amplifier and, through ten-ohm resistors, to *both* plus terminals of the amp. Make sure your amp can handle the strain of the extra two speakers, which must be identical to the original stereo speakers. Don't use four-ohm speakers, as these will overload the amp.

Dynaco offers a kit adaptor for connecting the extra speakers for about 20 bucks, which offers the additional convenience of switching the system out at any time you like. The separation available with this system is maximally about six decibels, even less than the other matrices, but it's a good, cheap way to get the effect of sound all around—and isn't that what quad's all about, anyway?

Chapter Seven

TEST-DRIVING A STEREO.

THE CHANCES ARE PRETTY GOOD that you don't know how to listen to a stereo.

I *don't* mean that you can't go sit down in your living room and find the volume knob (although, if you've got a volume knob and a hi-fi to go with it, why are you reading this book?). I'm talking about knowing what to listen for in evaluating the performance of a stereo system.

Most neophytes in the stereo game don't know what good sound is like. People who've had stereos for a few years have a somewhat better idea—they may be able to tell whether a system has good sound or not, but very likely they won't be able to tell you why.

A trained ear can tell very quickly where the strengths and weaknesses of a system lie, and getting a "trained ear" has very little to do with experience. The trick is to know what sounds to listen for, and what each sound means about the system performance. Comparing two systems in two different stores, both of which "sound okay," isn't going to help you make a decision. Being able to tell that one of them has good clean highs and crisp bass will.

The first thing to do is to choose a record from your collection, to take to the store with you. Stereo stores generally have a small collection of records to listen to, but chances are they've been used daily for the last six months, are scratched or worn past the point where they're any use in testing a system. There's another reason for bringing your own records—you *know* them. You have been listening to them for quite a while, you know where the bass player comes in and what the lead vocalist is saying—you can ignore the music and listen to the sound.

If your collection is fairly old, or your record player is an inexpensive model, it will be worth your while to purchase a new copy of an old favorite album for testing purposes. That way any distortion, noise or problem that shows up will definitely be the system's fault, and not the disc's.

Many people recommend using a classical recording as a test record, because they are usually made with greater care and there's no distortion or noise inherent in the music (who ever heard of a fuzzy oboe, for example?). If you're a fan of classical music, and you know what it's supposed to sound like, by all means. But if you're primarily into rock, use that; not only is it a better test of the system as you're planning to use it, but if you don't know what sounds right, you won't know what sounds wrong, either. Acoustic music is the best medium for testing a system with. There isn't so much going on that you become confused with which instrument is making which noise, and you can pick out one instrument to listen to and ignore the others. Also, acoustic instruments have distinctive sounds that don't resemble the sounds of distortion, as some amplified instruments do.

Rock'n'roll is often recorded as a solid wall of sound, with little differentiation between instruments (although modern recording technique is making for cleaner and cleaner records that still rock their asses off...). If you're a dedicated stomper, take along some quiet stuff for distortion checks *and* your favorite boogie band for power and bass tests.

When you get to the store, explain to the salesman that you want to test the system you're looking at with your own records. Make it clear that you want to operate the controls of the system yourself—*he* isn't the one buying the system, *you* are. If he appears reluctant, let him watch you for a while to prove you won't break anything, and when he sees you're competent to handle it he'll usually let you alone. By the way, make sure you *are* capable of running it without fouling anything up—show up straight and well-rested. If the music sounds good with a clear head, it'll sound good any other way you choose to be.

Quite often larger stores will have two big audition rooms—one for different speakers, one for amps and receivers—and a few systems scattered around the walls unplugged. Listening in the big rooms is fine for making preliminary choices of a system, but don't go out the door with your system until you've heard it as you bought it—same turntable, same receiver/amp, same speakers.

Here's why: The salesman may tell you that all amplifiers sound alike, so you can pick a pair of speakers just by listening to them. Then he'll take you into the amp room, where they've got a pair of $1000 speakers, and tell you that they are good enough to let you hear the differences between amplifiers. ("But you just said all amplifiers sound alike...")

To a limited extent what he's said is true, and you can weed out speakers or amps that are obviously unsuited to your needs. But there are differences between even the most similar systems—subtle and not-so-subtle—and you want to hear them before you choose. The amp you like may not have enough power to drive the speakers you like; there may be more noise in the $100 turntable you bought than in the $500 turntable in the testing room. You don't test-drive a car with a hopped-up engine, then buy the stock model, do you?

Another possibility but definitely second-best, is that the store may have a 30-day return policy if you don't like the system in your home. Best of all is to listen to the system in the store *and* have the 30-day trial period, because you may miss some deficiencies in a crowded stereo store.

If the salesman won't let you run the controls, if you can't hear the system as you're planning to buy it, if there's no return policy, if the system is out in the middle of the floor and the madding crowd is noisily flowing by—thank the man and split. *You* are buying a stereo. If he doesn't want your business...brother, that's not your problem.

It's pretty unlikely that the store is out to stick you with a turkey. Stores like that don't stay in business long. It is simply the March of Modern Marketing—getting you in quick, through quick, and over to the cash register is more convenient for the store...and most American buyers have been pushed around for so long they don't even realize that it's happening. But it's *your* money, so make it clear that you're calling the shots.

Okay. Now let's suppose that, instead of giving you a lot of gas like I've detailed on the last few pages, the salesman says, "Sure. Go ahead." Now what?

Don't turn the power on yet. Set the tone controls flat, dead center. If there is a tone control defeat switch (which cuts the tone controls completely out of circuit), set it to *defeat*. Loudness control, high and low filters, tape monitor switches should be OFF. Last, set the volume control to about nine or ten o'clock.

Now turn the power on. Did you hear a click or a thump? That's the *starting transient* caused by the power supply charging up to full voltage. If it's overly loud or high-pitched, it could damage your speakers, especially with the higher-powered amps on the market. Turning the power *off* for about ten seconds may produce another transient. Some amps have transient suppression circuits built in, so if you hear nothing don't be alarmed.

If you do hear loud clicks or thumps, try turning the set on and off a couple of times with the volume all the way down. The transients should disappear—if not, don't buy the set. If they do disappear, and you eventually buy the set, remember to turn the volume down before you turn it on and off—you'll save wear and tear on amp *and* speakers.

Once you're satisfied transients won't be a problem, start listening to the set. Not the music yet, just the set; turn the volume up about halfway and listen for hiss, buzzes or hum. If you can hear any at this setting, you'll probably hear it between bands of a record, or behind quiet passages in the music.

The problem may lie in poor cabling by the store—hum and noise are very easy to introduce in connecting cables. If the set is noisy, have the salesman disconnect cables one by one to see if the noise goes away. (Don't you do it, at least not without permission—it's still his set.)

If the noise does go away, then either the cable or the equipment at the cable's other end is faulty. Ask the salesman to get it repaired before you come to listen again. If the noise does *not* go away, it's in the amp (or being picked up by the amp, at least), and that's a bad sign.

Sometimes simply *moving* the cables without disconnecting them will change the noise level. That indicates the noise is coming from the cables, probably through bad grounding. Cables—and components—can also pick up signals from radio stations, fluorescent lights, light dimmers and so on. The store should have set the system up as cleanly as possible. If you keep running into noises, and the store can't seem to cure them, you might consider finding a store that knows more about what it sells. Of course, it could simply be that the equipment isn't any good.

Here's what to listen for:

HUM: A low note in the deep bass, or sometimes a harsh raspy buzz (caused by harmonics of the 60 Hz fundamental hum note). Usually caused by poor grounding or shielding—if playing with the cables won't make it go away, don't buy the amp.

WHITE NOISE: A trebly hiss or rushing sound, like the noise of a waterfall. Caused by electrons rattling around randomly in atoms. With proper design this should be nearly inaudible at listening levels—if it's loud enough to hear over a conversation when the volume knob is all the way up, you've probably got too much noise in the amp. Find a better amp.

POPCORN NOISE: Intermittent pops, clicks or crackles. May be caused by circuitry, or by faulty cable connections. Shake the cables to see if they're causing it—if not, it's the amp. If so, replace the cables and keep listening.

It's possible for a faulty component to get past the inspectors at the factory. It's also pretty likely that the demo system in the showroom has had some pretty hard use, been subject to vibration, cigarette smoke and Coca-Cola, so don't give up if the first amp sounds bad. It may have just broken down.

Now it's time for the music. Put your album on the turntable, turn the volume up to your usual listening level and just a little beyond, and listen:

BASS: Listen to bass guitar, organ. Synthesizer bass is not so good for testing bass response, as it usually contains harmonics up into the midrange. Same thing with fuzz bass—too much distortion to be a useful guide. Bass *drums* are also pretty useless for testing bass response. The reason is that record producers get rid of the bass *thump* and emphasize the midrange *whap* of the bass-drum beater, so that the sound fits more easily into a record groove.

If the bass sounds like fuzz bass (and you know it isn't), try turning the volume down slightly. If the fuzziness disappears, the amp does not have enough power—the fuzziness was caused by overdriving the output of the amp. Another possibility is that you're overdriving the *speaker*, and it's distorting the bass note. Try another pair of speakers with the same amp.

Are the lowest notes of the bass equally loud with the higher notes? Poor low frequency response can make it sound like the bass player forgot to hit a note when he was supposed to. Listen also to see if only the very lowest bass notes fuzz out—that's in the speaker, and it's called *doubling*; the speaker is vibrating at twice the frequency you're feeding it. The cure, sorry to say, is a better speaker.

At listening level, turn up the bass control until you're satisfied with the tone quality. Now listen again for fuzziness or distortion. If there isn't any, fine; if it sounds good with tone controls flat but distorts with the bass turned up, you probably need more power, or a more efficient pair of speakers. Extra bass needs extra power.

Now watch the pilot lights, the dial light and the meter while you play a good bass passage. Do the lights dim or flicker? Listen closely just *after* each note is played. Does the volume drop slightly for a split-second, or is there a moment of hum or buzz? These are all symptoms of poor power-supply design, and a tipoff that this is *not* the amp for you.

MIDRANGE: Listen to guitars, lead vocals, pianos, flutes. The piano and the human voice are two of the most difficult sources to reproduce accurately, because they are full of sharp attacks, clicks, buzzes and tones arranged in a complex pattern. Any distortion of that pattern, however slight, shows up clearly to people who know what the real thing sounds like.

Voices should be clear and easily understood (make sure that the *recording* is good. Rock 'n' roll vocals are often quite muddy. Folk vocalists, especially women, are usually recorded very cleanly). If the voices or piano notes sound raspy, they are being distorted. Turn down the volume slightly—if the raspiness disappears, it's the speaker's fault, if not, the amp's. Listen to voices in both loud and soft passages, accompanied and solo. Often an amp that handles solo voice splendidly is useless for band-and-vocalist.

Raspy sound may also be the fault of the turntable. Turn the volume all the way down, lift the tone arm off the record and check for dust on the stylus and on the record. If the stylus is dirty, blow the wad of dirt gently away. NEVER flick it away with your finger—you're too likely to take the diamond with the dirt. Check to see that the tracking force and anti-skating adjustments are correctly set—over- or under-pressure can foul up the sound and destroy the record. In the final resort, try a different model turntable. Especially in lower-priced systems, the turntable may be introducing more distortion than amp and speakers combined.

There are controls on the back of most speaker systems for controlling the relative level of the midrange and tweeter speakers against the woofer. If the music sounds overly shrill or hollow, try adjusting these for best sound with the amp tone controls dead flat. Quite often stores will set these incorrectly, usually unintentionally. Some stores try to steer you away from low-profit brands by misadjusting the controls deliberately. Either way, you're not getting a fair test of the speaker's performance. Just make sure you don't turn the tweeter control up too high, or you'll burn it out on the first Jimi Hendrix lead.

HIGHS: Listen to human voices, violins, tambourines, cymbals. The splash of the cymbal should be clean and sharp. If it sounds like somebody draped a rag over the cymbal then the tweeter is poor, or else it's broken. Tweeters in stereo stores are quite often blown out by demonstrations and occasionally the salesman doesn't realize it. If the highs sound raspy or harsh, try switching back and forth between the left and right speakers to see if there's a difference (do this in mono). It's unlikely *both* tweeters will be fried, and you'll get a fair idea which one is working properly.

Clicks ("K," "t") and sibilants ("s," "z") in the human voice are good checks of high-frequency performance. Sibilants should not sound whistly, nor like someone was spitting into the microphone. Attacks should be crisp and clear without making a thump. If it sounds like someone is spitting into the microphone or hissing in your ear, the highs are distorting. Turn down the volume, or the treble control—if the distortion goes away, it's in the amp. Chances are it will *not* go away—good tweeters are tricky to design, and they don't usually show up in speakers selling for less than $100 apiece.

For a final check on the system, set the filters, tone controls and so on to settings that sound good to you, and slowly turn the volume up until you begin to detect distortion, raspiness, poor sound or what have you. This is as loud as you're going to get with this system. If it's satisfactory, you're in.

A word of caution—after four or five of these sessions you're going to get a little bored comparing systems (that's another reason for getting a full night's sleep before checking out a stereo). All that I can say is that your stereo is going to be with you for a long time, and it's worth your while to spend a little extra time finding the best system you can.

Chapter Eight

FEEL THE POWER.

THE HEART OF YOUR SYSTEM IS THE AMPLIFIER, and you can have it in three different configurations. If your tuner section, preamplifier and power amplifier are all in the same box, it's called a *receiver*. Separate tuner but preamp and power amp in one package makes it an *integrated amplifier* (no relation to "integrated circuit," which is a lot of transistors and other goodies hooked together on a single near-microscopic chip of silicon). *Preamplifier* and *power amplifier* can also be separately housed.

There are advantages and disadvantages to each approach. If everything is in one box, you only pay for a single power supply (instead of three) and a single cabinet. On the other hand, if you decide to trade up to a better system, you'll have to turn in a receiver and start from scratch; with separate components you may trade up only the component you're not satisfied with.

On the whole, receivers are more popular and less expensive; separate components in the same power range are slightly more expensive but a little more reliable and offer somewhat superior performance. Separate components have to stand on their own merits, while a receiver with merely adequate tuner and preamp sections can be sold on its power performance. The catch with separate components is that few, if any, are offered in the low-power, medium-performance range; most separates are high-end equipment.

The preamplifier is often called a *control amplifier*, which is somewhat more descriptive of its purpose: It adjusts volume, tone and balance, selects the proper input, routes signals to and from tape decks and outboard accessories, and generally keeps track of the audio for you. There isn't a great deal of amplifying going on inside—the input signals come out at roughly the same volume at which they went in, although most preamps have a high-gain circuit called a *phono preamplifier* built in, to raise the millivolt signals from the turntable to a level where the control amplifier can deal with them.

Choosing a preamplifier is simple: Just make sure that the controls offer enough flexibility for your use, then check to see if the specs are adequate, then see if you can afford it. Don't skimp on flexibility—you don't have a tape deck now, but you might get one later, and there's no sense in having to buy a new preamp so you can use it. Most control amps have a *tape monitor* switch which allows you to monitor the signal being recorded while it's taping, to insure it's going on tape correctly. Some systems have two, which is more than enough for most of us.

Loudness is a boost of low and high frequencies at low volume, to compensate for the human ear's reduced sensitivity to bass and treble in quiet sounds. The better designs automatically disconnect the loudness as you turn the volume up; the less expensive leave it to you to switch out the boost. Some inexpensive designs boost only the bass—listen before you buy to make sure the treble is increased when the switch is cut in.

If you're a purist, you'll want a switch to defeat the tone control circuitry, which adds a very slight amount of distortion even with the controls in dead center. In most cases the added distortion is negligible compared to the performance of the entire set.

Tone controls boost or cut the bass and treble ranges of the music. Most controls are continuously variable; some manufacturers have gone to stepped controls which add a discrete amount of boost or cut at a given frequency. Aside from the extra reliability of switched (stepped) controls over variable ones, the difference is mostly cosmetic; few of us can hear such fine tonal distinctions anyway.

Some designers have added a midrange control as well, or even *graphic equalization* to individually boost or cut several ranges of frequencies across the audio spectrum. This can be very helpful in correcting the sound of a boomy room, or spiffing up a poor recording. For those whose amps don't have graphic EQ, outboard equalizers are available from several makers which connect to the tape monitor circuitry for easy switching.

Low and high filters (or rumble and scratch filters) sharply cut out the very deepest bass and highest treble, to improve the sound of a scratchy record or a cheap turntable. They chop out some of the music, too, but they're real handy to have when your weird friend comes over with his collection of old 78's and 45's.

The control circuitry leaves the signal with the desired response, but without the power needed to push a speaker cone around. Thus it is sent to power stages after preamplification, for beefing up.

Separate power amplifiers tend to be designed for the higher-power ranges, and for better distortion performance than the average receiver. This allows the owner to choose the desired level of fidelity in his control circuitry separately from the power requirements. With most receivers and integrated amplifiers, increased fidelity comes only with increased power, and you pay for both.

Separate power amplifiers have the added advantage of keeping the high-temperature power circuitry separate from the more delicate low-level circuitry, and isolating more circuitry from damage in case of breakdown. To its disadvantage, the separate power amp is big, usually bulky and dull-looking, heavy and hard to hide (at least, with adequate ventilation), and expensive.

Those who don't want a full receiver but can't afford or don't choose the separate power-amp system may try the *integrated amplifier*, which combines control and power circuitry—but without tuner circuits—in a single component. Integrated amplifiers are available at almost all power levels except the super-power, with performance matching receiver specs in a comparable price range.

Integrated amplifiers are available at power and performance levels matching receivers, but at prices $100 to $200 less for comparable performance. If you don't mind giving up FM for a while, you can build an excellent system with turntable and integrated amp alone, adding the tuner later. This may even save you money in the long run, because most broadcast stations have no better fidelity than they absolutely have to. Rather than spending lots for excellent tuner circuitry in a receiver, just to gain high-performance control and power circuitry, you can get an excellent integrated amp and a medium-performance tuner. Of course, if you're later blessed with a broadcaster who considers radio a joy instead of just a job, and keeps the quality high, you can trade the tuner up without sacrificing the amplifier.

Your power requirements will vary directly with the desire for clean sound and inversely with what you can afford (a friend of mine also claims that, the younger you are, the more power you need, which explains those teenagers pawning their kid brothers for a megawatt monster system...). Like everything else, listen and make your own decision, but plan ahead. Your tastes and ear will change as you listen to your system, and you may need more power later, so leave some margin if you can. Skimping now means living with it later—or paying for it, and usually paying more than you would have earlier.

Chapter Nine

SPINNING THE BIG WHEEL.

A good disc-playing device lies at the heart of a quality system. Properly chosen (and maintained), a good record player will bring you every sweet and delicate note your records hold. Get a poor one, and not only will you hear less of the music, but chances are good that the stylus will scrape the music right off the walls of the groove.

Disc-players fall into two categories. An *automatic changer* plays a stack of records one after the other, and shuts itself off at the end of the stack (it can also be set to repeat the last side indefinitely). *Manual turntables* just go round and round, leaving it to you to stir your butt and flip the LP after it's over.

Over the years, changers have been overwhelmingly more popular than straight manual turntables—a slightly misleading statistic, since many changer owners will not play a stack of records on their changers, but use them for single plays with the convenience of auto-shutoff.

Very recently the manufacturers have finally realized this fact, and some have come out with a hybrid, an *automatic turntable* that shuts itself off at the end of the side. The fancier models also set the needle down at the beginning of the side, and return the tone arm to its rest after the record is over.

Since the beginning of high fidelity there has been a battle between patzers and purists over which system is "better." Purists maintain that the linkages used to sense the end of a record and to move the tone arm put an uneven strain on the moving parts and cause the music to become distorted or off-speed. Listeners with a less fanatical devotion to the state of the art usually opt for the convenience instead.

Modern changers will play records every bit as accurately and delicately as a manual turntable. The chief difference between the two types lies in their cost; for equally good performance, the extra linkages and control levers of the changer add up to higher price...as much as twice the cost of the comparable manual. Manufacturers who claim to make the extra linkages at no extra cost are kidding themselves and you.

Curiously, the least expensive (read, "cheapest") changers on the market cost less than the least expensive manuals. How do they do it? Simple—*really* poor quality. A decent changer—one that will produce good quality sound and won't ruin your records—will cost at least $100, and a *good* changer goes for $150 or better. In contrast, there are several good manual turntables available (with tone arm) for around $100.

One further irony: The *most* expensive disc-players are manual turntables, and the manufacturers don't even supply a tone arm as standard equipment. Reason? The only people who buy these marvelous machines are the dyed-in-the-wool fanatics, and they usually have near-religious convictions about the "best" tone-arm (and the "best" turntable, and the "best" amp, and the "best" brand of toothpaste...).

So how do you tell good from bad? I thought you'd never ask...

The most obvious feature of any disc-player is the platter—a circular chunk of metal that goes around and around and around and around, at a steady rate of 33.33 revolutions per minute, or at 45.00 r.p.m. (For you history buffs, there are also records—and record players—at 78.26 r.p.m., and 16.66 r.p.m. Unless you're a Rudy Vallee freak, these speeds probably hold little interest for you.)

The accuracy with which these speeds are maintained is a measure of the quality of the system. If the speed is *not* constant, but varies slightly as the platter revolves, the music will slide up and down in pitch like a siren.

Try an experiment. Take a 45-rpm single, place it slightly off center on your turntable, then play it. The needle is traveling in the groove more slowly as it nears the center, faster the farther out it goes. The rising/falling effect is called *wow*: A rhythmic variation of speed as the turntable goes around.

Now let's suppose that the motor in the player jumps up and down slightly in speed, say 20 or 30 times a second. This causes a vibrato effect in the music, like they were singing the opera while jogging around the stage. This is called *flutter*—a relatively high-frequency change in platter speed.

Making the platter heavier helps both of these problems—the inertia of the heavier platter acts as a flywheel to smooth out speed variations. It also places a greater load on the driving mechanism—hence, more durable (and expensive) parts are required.

As the turntable gets older, parts wear and get loose, and the axis of the platter, instead of staying precisely vertical, may wobble from side to side, and the surface of the platter moves up and down slightly. There's no way the needle can tell the difference between this and a groove that moves up and down—the cartridge sends this new signal to the amp, and you hear it as a low, grumbling bass note called *rumble*. Rumble can also be caused by irregularities in the surface of the platter and of the record disc. In fact, in modern players above $250 or so, the major cause of rumble may be in the disc alone—the mechanical rumble has been reduced to inaudibility.

Rumble, wow, and flutter are measured in *percentages* of the full output signal from the phono cartridge: The lower the percentage, the less undesired output and the better the system.

The platter drive mechanism has an effect on how good the wow, flutter and rumble specs will be. There are any number of ways to make a motor spinning at 1800 rpm drive a platter at 33⅓, but the two most popular are *idler drive* and *belt drive*.

In idler drives, a rubber wheel is placed between the shaft of the motor and the platter rim—usually underneath the platter and against the inside rim (diagram). The motor spins the rubber *idler wheel*, which turns at a fraction of the motor speed and transfers energy to the inside of the platter rim, which then spins slower again.

Idler drive is nothing more than a gear train, with friction from the rubber replacing the teeth on the gears. The idler is used because the flexible rubber tends to soak up vibrations and slight speed changes without passing them on to the platter itself.

Belt drive transfers energy from motor to platter through a rubber band looped around the motor shaft and the platter. Some designs loop the belt around the outside of the platter, others use a smaller special rim built into the underside of the platter, to hide the belt from dust, moisture and curious kittens.

Belt drive usually gives superior isolation from the motor vibrations, and has been preferred for manual turntables. The long unsupported span of rubber isolates vibration very nicely. Unfortunately, it also stretches when the motor is turned on and off and the platter is getting up (or down) speed. Purists usually don't mind the extra second or two, but changer mechanisms put too much strain on the belt, and almost all changers on the market use the idler drive system (though there is one manufacturer making a belt-drive changer).

A third drive method, *direct drive*, has recently been developed. In this system, the platter is part of the motor itself—coils of wire are wound on the platter, and complementary coils are mounted below the first ones. When the proper current is applied to the two sets of coils, the platter turns from the magnetic force between the two coils. The advantage is that the platter has no mechanical connection at all to the driving force—the magnetic fields "push" the platter without touching it. The only mechanical connection to the platter is through the bearing on which it spins, and that's easy to make rumble-free. Direct drive platters approach utter silence in their performance, and you can get either manual or automatic models. The only hassle is that winding coils balanced for weight and magnetism is tricky and expensive—direct-drive is *not* your bargain of the month.

Now that we have the platter going around more or less steadily, let's look to the *tone arm*. The sole purpose of the tone arm is to hold the phono cartridge in the proper place for playing the record. The difficulty lies in making the record think that the tone arm isn't there at all. Tone arms are hinged at their base to move side-to-side (to follow the needle into the center of the record) and up-and-down (to follow the warps and irregularities in the record surface). These hinges, or *gimbals*, must not bind, or the stylus will be dragged against the sides or bottom of the groove—or float above the groove as the record falls out from under the needle and the arm doesn't drop with it. At the same time that the arm is allowing free movement of the stylus, it must also hold the stylus in close contact with the groove, or you'll lose high frequencies and possibly damage the record.

The tone arm makes this compromise by keeping the gimbal movement free and letting gravity hold the stylus in place. By counterbalancing the cartridge with a weight at the other end of the arm, the stylus pressure can be adjusted down to fractions of a gram. Even this small *tracking pressure* can cause problems, as we'll see later.

You'll also find an adjustment at the tone arm base for *anti-skating*. Early cartridges required tracking pressures of multiple grams—some even measured their tracking in ounces! Skating force was swamped out by the other, more serious problems caused by the high weight, but as gimbals and cartridges got better and lighter, audiophiles began hearing a slight raspiness in their records. Finally the math boys sat down and figured out what the trouble was.

In theory, a line through the center of the cartridge should be like a *tangent* to the circle of the record groove, perpendicular to a line between stylus and center post. Unfortunately, as the tone arm swings, so does the cartridge, and tangency is lost.

As the groove moves under the needle it tends to pull the tone arm towards the center of the platter. This is the *skating* force, which you can see demonstrated by "playing" a grooveless record and watching the needle skate inwards. Some stereo test records have a blank band or side for just this test.

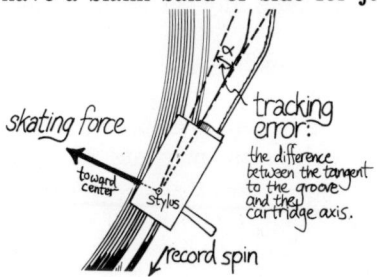

skating force — toward center — stylus — record spin

tracking error: the difference between the tangent to the groove and the cartridge axis.

The skating force presses the stylus more strongly against one side of the groove than the other, distorting the playback from both sides slightly and possibly damaging the inside wall of the groove in severe cases. The heavier the tracking pressure, the greater the pull towards the center. Worse, the skating force changes as the needle gets closer to center and the tangency error gets smaller.

Various schemes have been devised for putting equal pressure on the inside of the tone arm to compensate for the tracking force, ranging from ingenious counterweights to complicated strings and springs and slides and suspended weights. The very best arms usually use strings and weights, since they compensate more exactly. The disadvantage of the hanging weight system is that it attracts kittens and children (one cat I knew loved to ride the platter around and bat at the weight every time it went by...).

Even the very best turntable/tone arm combination is useless without a good cartridge. There are two major types of phono cartridges in use today, the *ceramic/crystal* and the *magnetic reluctance* types. Of the two, the magnetic is far more sensitive and high-fidelity.

Certain types of crystal and man-made ceramics have the property of generating a voltage when squeezed or pulled apart. This *piezoelectric* effect, from the Greek word for pressure (*piezo*), allows a manufacturer to hook a needle onto one end of the crystal and let the groove move the needle and squeeze the cartridge.

The trouble is, the forces needed to get a decent voltage from the cartridge are too great—you scrape the music right off the groove walls; and crystal cartridge response isn't high-fidelity, either. Since they're cheap, durable, and require less amplification than magnetic cartridges, manufacturers use crystal or ceramic cartridges in kids' phonographs and stereo consoles for people who don't know any better. Don't let anyone sell a crystal cartridge to *you*.

Magnetic cartridges use a different principle: that a wire moved through a stationary magnetic field, or a stationary wire in a changing magnetic field, will have a current induced in it. Next step: hook a couple of tiny magnets, or some small wire coils, to a metal tube with a diamond stylus on the other end, and you've got a cartridge. You can make the metal tube, the stylus, and the coils or magnets as small as you like, for extra delicacy in playing the record.

Of course, at these low pressures the voltage you get from the cartridge is very small, and must be *preamplified* before it's large enough to use in an amp. The extra amplification makes the phono input more sensitive to noise and hum pickup as well, which is why most players have an extra ground wire to hook up, and more or less elaborate shielding of the wires to the cartridge.

Even at tracking pressure of a few grams it's possible to damage the record. The stylus is a tiny chunk of diamond, a fraction of a thousandth of an inch in diameter (one one-thousandth of an inch is called a *mil*). The cross-section of a standard *conical* stylus, at the point where it touches the record, is a circle 7/10th of a mil in diameter. The more expensive *elliptical* stylus is even smaller—its cross-section is an ellipse

7/10ths of a mil by 3/10ths. The sharper "corners" of an elliptical stylus can follow the groove into smaller variations than the conical stylus, resulting in better high-frequency tracking.

Now, remember that these styli are touching the sides of the groove over a very tiny area—less than one one-millionth of a square inch. One gram of pressure on such a small area results in *tons* of pressure per square inch! That's more than enough to deform the vinyl for a split-second as the needle goes past, or to damage it permanently if the pressure is too great. This is why the cartridge maker specifies a maximum tracking pressure for his stylus.

It's also the reason for letting a record sit for an hour after you've played it before playing it again. Some plastics, especially vinyl, have a "memory" for shape, and will "recover" after an hour or so. Replaying the same groove over and over again without letting the vinyl recover will permanently deform the groove.

The introduction of CD-4 quad produced requirements for high-frequency tracking at twice the pitch the human ear can hear. To let styli track this without damaging the groove, a Japanese engineer named Shibata and his team developed a stylus shape with even smaller "corners," but which touched the groove over a greater *vertical* range, so that the overall pressure was no greater. Similar stylus shapes have been worked out in the U.S. as well.

Records are not made with a flat frequency response—on purpose. Dirt, dust, static and other record noise shows up as high frequencies in the playback. By boosting the highs before the music goes in the grooves, the treble range tends to drown out the high-pitched noise. A complementary circuit in your preamplifier drops the high frequencies back where they belong, and drops the noise down to inaudibility with them. This boost and cut technique is call the *RIAA curve*. The Recording Industries Association of America uses a standard equalization on every record made by its members (which is to say, nearly every record maker around). This allows a single circuit (built into the preamplifier in your system) to compensate for all records.

Working with component sizes of a thousandth of an inch and pressures less than the weight of a feather, you can see the need for delicate, precision equipment. Anything less will give you poor music and possibly ruin your records—and *that's* why the cheap changers aren't worth the time to drag home.

Chapter Ten

THEY FLOAT THROUGH THE AIR WITH THE GREATEST OF EASE.

It's not uncommon for a hi-fi to contain no turntable at all. Most hi-fi owners spend more time listening off-air than off-record, and some people get along fine with just a receiver and a deck for taping off the air.

Radio comes in two flavors, AM and FM. AM broadcasting has been going on since KDKA set up shop in Pittsburg in the early 1920's. Many people think FM is a relative newcomer to broadcasting, but it was invented in the early 1930's by Colonel Edwin Armstrong who was granted the first commercial FM broadcast license in Hartford, Connecticut a few years later.

AM stands for *amplitude modulation*, which means the strength ("amplitude") of the transmitted signal varies with the musical signal put into it. In *frequency modulation*, the strength of the signal remains constant but the frequency the transmitter is tuned to moves up and down in time with the music.

Colonel Armstrong must have been a little lonely in those early years. Very few FM stations were licensed until after the war, most of these to owners of AM stations in the same city, who broadcast the same material over both facilities ("simulcasting"). It wasn't until the early sixties that FM began to come into its own, for three separate reasons.

First, the FCC required stations in most cities to broadcast separate programs at least 50 per cent of the time, cutting down on simulcasting. Second, stereo transmission was developed and FCC-approved. And last, the hi-fi boom and rapidly-developing technology had placed inexpensive FM in enough homes to find an audience. Even so, a study as late as 1970 showed the average FM station losing money and supporting itself on money from the associated AM broadcaster.

With the exception of stereo and SCA background music, the technical standards for FM have changed very little since the 1930's. The 88 to 108 megacycle band (they called 'em cycles back in those days) was considered a throwaway—far too high in frequency to be of much use in long-distance communication. So they could afford to grant privileges like a 50 to 15,000 cps frequency response; it was a far wider response than music "needed," but since no one can hear the signal anyway, why not?

If someone had told the Federal engineers that we'd be broadcasting pictures at ten times the frequency and sending radio signals—and men—to the moon less than forty years later, they would have thought he was crazy. There was, of course, no way that the FCC could have foreseen the growth in technology brought on by the war and by the postwar economic boom. Still we now find ourselves with eminently practical means of transmitting and receiving FM signals but limited to the standards set in the 1930's.

The modern FM tuner is capable of performance some of those old-timers would have considered impossible. The specs for this performance, alas, have most people just as confused as the engineers, so let us go on a guided tour of a tuner spec sheet.

The strength of the FM station's signal at your tuner's antenna terminals may be as tiny as a millionth of a volt or less. The tuner's ability to respond to such a small signal is the measure of its *sensitivity*. One of the peculiar effects of FM is that as the signal gets stronger the music does not get louder. What *does* happen is that the background noise gets quieter. This *quieting* effect provides a way to measure sensitivity—if, for example, two microvolts (millionths of a volt) of signal are required for the music to be 20 decibels louder than the background hiss, that tuner is said to have a sensitivity of "two uv. for 20 db. quieting."

The 20 decibel quieting figure is a standard one in the industry. The Institute of High Fidelity, a group of hi-fi manufacturers, uses it as their standard, so that you might see a sensitivity rating of "two uv. IHF" for the above tuner. The two specs mean exactly the same thing.

As the signal gets stronger still, the noise should drop below 60 decibels quieting, below the level at which the human ear is bothered by it. Once the signal voltage reaches that level, any noise you might hear comes from the station itself and not from the tuner (at least not from a properly operating tuner . . .).

In addition to picking up the station you want to hear, the tuner is also rejecting the stations you *don't* want to hear—that is, all the other FM stations that you can pick up. The measure of the tuner's ability to reject these unwanted signals is its *selectivity*.

The FCC assigns FM channels every 200 kilohertz across the FM band, at odd decimals (88.1 Megahertz, 88.3, 88.5 etcetera, up to 107.9 Mhz). Selectivity is measured by tuning to a signal at 88.1 Mhz, for example, and measuring how well an 88.3 Mhz signal is rejected. The strength of the unwanted signal is measured as a decibel ratio below the wanted signal. Here again, 60 decibels is an adequate rating—higher numbers indicate even better selectivity.

Suppose you're listening to a very weak, distant station, but you're living near a strong station that you *don't* want to listen to. If the nearby station is strong enough, it may transfer some of its modulation (that is, the music) to the weaker station's signal inside your radio. The process is called *cross-modulation*, and it's obviously a no-no. The effect sounds like both stations are playing at once.

Modern circuitry is pretty much immune to cross-modulation effects (unless you happen to be living next door to somebody's FM station); often enough cross-modulation is not even listed in a tuner's spec sheets. When it is, you'll find it listed in decibels again; same trick here—the higher the number of decibels, the better the rejection of cross-modulation.

What happens if you live halfway between two FM stations that have the same frequency? Here we find another curious property of FM—the *capture effect*. If two signals are on the same frequency but one is sufficiently stronger than the other, you'll hear only the strong one. In effect, the tuner is "captured" by the stronger station. The difference in strength required for this effect (measured in decibels again) is the *capture ratio* of the tuner. Typical capture ratios are two decibels or so. Two decibels corresponds to about a 50% difference in signal strength—not very much in radio. The *smaller* the number of decibels, the better the capture ratio.

If you happen to be "captured" by the wrong station, just get a directional FM or TV antenna and point it at the station you want to hear. Or, point it *away* from the station that you *don't* want to hear. If you're on a line between the two stations, do both.

The trick is to orient the antenna so that you get best reception on *all* the stations you want to hear, and the only way to do that is by trial and error.

The circuitry that turns the FM signal into an audio signal that your amp can process works best when tuned to a single frequency. Most receivers translate the incoming FM signal down to this *intermediate frequency* and then do their stuff (this type of radio, by the bye, is called a "superheterodyne," and was invented back in the early 30's by our old friend Colonel Armstrong).

Of course, if the set happens to pick up an outside signal at this intermediate frequency (or "IF"), it won't be able to tell the difference between the outside signal and the translated FM signal. How well the set is shielded from this outside pickup is its *IF rejection*. Decibels again, at least 60 or more.

The interference doesn't have to come from outside... there are enough oscillators, mixers and assorted other sources inside the tuner itself to generate a spurious signal. These *spurious responses* should also be 60 db or more below the audio output.

One of FM's greatest advantages of AM is its immunity to impulse noise—the kind of electrical noise caused by spark plugs, lightning, and various other electrical disturbances. These signals vary in strength; they are *amplitude modulated* (like AM broadcast signals, but at FM frequencies). The tuner should reject AM signal components—how well it does so (in decibels) is its *AM rejection*—50 decibels is a minimum, more is gooder.

AM rejection is *not*, as some people think, the set's rejection of signals on the regular AM broadcast band. If you're picking up an AM broadcaster in your FM listening, chances are there's a loose wire or a faulty plug somewhere in your setup. The FM tuner is very seldom the culprit.

It's possible, with an extremely strong AM broadcast signal, for the *audio* stages of the tuner or the amp to be picking it up. Again, the wiring may be defective, but even clean wiring can pick up a strong signal.

Here's a quick test: If you hear the AM station with the amp switched to "phono," or the interference doesn't change when you turn down the volume, it's definitely *not* in the tuner. If you can't get rid of it by playing with the wiring, try moving the set to another spot in the room, or another room. Sometimes moving it three feet will do it!

If you can't get rid of the problem no matter what you try,

there's help available. Most service shops aren't very familiar with the proper procedures for curing RF pickup (yup, that's what your problem is). Get in touch with a local ham radio operator or one of the city's ham clubs. (You may have to put up with a little missionary work trying to get you to become a ham.) These clubs are usually all too familiar with interference, and may know just how—or who—to straighten you out. You should also try the Chief Engineer of the station that's interfering with you.

The FCC licenses some broadcasters to carry background music or other signals on a special subchannel of their frequency. This *Subsidiary Communications Authorization* (or *SCA*) signal is decoded in special receivers at your local department store, who play it in the ceiling. You shouldn't hear it at all—how much you don't hear it is the *SCA rejection*—60 decibels or better, just like the others...

In the early 1960's the FCC authorized a system of *multiplex* broadcasting. The word "multiplex" does not mean "stereo," as many people think. All it means is two or more channels carried on the same signal. And here's how it's done:

To maintain *compatibility* (that is, to make sure that the mono listener gets both channels of the stereo signal, in mono, on his speaker), we take the left channel, A, and add it to the right channel, B. This A + B *sum channel* is what the mono listener hears.

We also generate a *difference* signal, A - B. This difference signal is encoded onto a *subcarrier* at 38 khz, far above the range of human hearing. In the encoding process, the 38 khz subcarrier is cancelled out, leaving two *sidebands*, each fifteen kilohertz wide, from 23 khz to 38 khz, and from 38 to 53 khz. These sidebands are still above human hearing, which cuts off around 20 khz.

There is also a *pilot* carrier at 19 khz. If we broadcast the two A-B sidebands with the 38 khz subcarrier, the subcarrier would interfere with the sidebands in reception and produce audible results. But we need the 38 khz carrier to decode the A-B signal. So before it's eliminated in the encoder, we divide the 38 khz frequency in half, and send out the resulting 19 khz signal with the rest of the mess.

In the receiver, the 19 khz pilot is doubled back to 38 khz, used to decode the A-B signal, and we're in business. Now all we (meaning the tuner) have to do is add and subtract our sum and difference channels:

Making the Connection

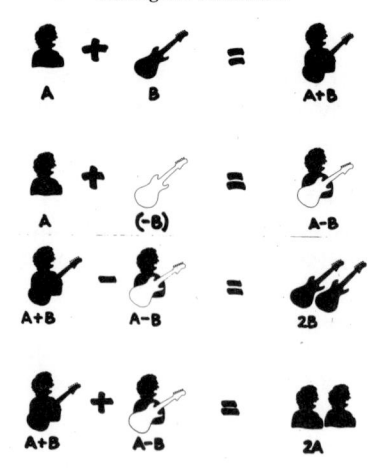

👤 + 🎸 = 🎸
A   B   A+B

👤 + 🎸 = 🎸
A   (-B)   A-B

🎸 − 🎸 = 🎸🎸
A+B   A-B   2B

🎸 + 🎸 = 👤👤
A+B   A-B   2A

$(A+B) + (A-B) = A+B+A-B = 2 \times A$, which is our left channel,
and
$(A+B)\ (A-B) = A+B-A+B = 2 \times B$, which is our right channel.

At this point you should have a few questions. Yes, the 19 khz pilot tone *is* in the range of human hearing—that's one of the unfortunate drawbacks of the system that the FCC decided to overlook.

Yes, it's complicated—but no more so than two or three other schemes the FCC considered before choosing this one (some of which were technically superior, we might add ... *Ed.*). The problem is very similar to broadcasting a TV signal that's black-and-white for some folks and color for others—a problem they solved in a very similar way, too.

Yes, it works out nicely on paper, but I can testify from experience that it's a bitch to adjust properly, receiver *or* transmitter! Of course, there's all them technicians at the factory who are supposed to adjust your set before it gets to you. Now we can go back to the spec sheet and see how well they did.

The 38 khz subcarrier is above the range of human hearing, but some of the intermodulation products between it and various other signals inside the radio are not. Therefore the tuner suppresses the subcarrier before it can do any audible damage—the *subcarrier suppression* should be at least 35 decibels or more.

The 19 khz pilot *is* audible, and also causes intermodulation garbage. So we suppress this, too; *pilot suppression* should be 35 decibels or more.

Putting two channels on one FM signal is like pouring red and green paint down the same pipe and then separating them at the other end. How much red there is in the green paint—that is, how much left channel there is in the right channel signal, and vice versa—measures the *channel separation*. There should be at least 30 decibels of channel separation; the leakage from one channel should be 30 decibels or more *softer* than the signal in the other channel. FCC regs specify 29.7 db. as a legal minimum for transmission; receivers should be as good or better.

Incidentally, the FCC has not yet settled on a technique for broadcasting multiplex *quad*. Any system they approve would have to give a mono signal to mono listeners, a good stereo signal that sounds natural in stereo, and fair separation (on the order of 25 or 30 decibels) for four separate quad channels. That, my friends, is a tall order, and all the systems so far proposed either don't hack it, work under the most optimum of conditions, or don't meet some other requirement for an FM signal.

Quite a few FM stations claim to be broadcasting in quad; what they are doing is to broadcast a stereo-encoded *matrix* quad signal—in stereo—for people with quad decoders at home. Worse luck, much of what makes multiplex stereo work fouls up the matrix decoding process. When you pour four colors of paint down the tube, so far you get only mud at the other end.

One of the worst problems with quad is *multipath distortion*, where the signal from the FM station is bounced off a building or a mountain on its way to your antenna. If you receive both the direct wave and a reflected wave which has traveled slightly further, and is therefore a little out of step with the first, they will interfere with each other.

The result may be a noisy stereo signal, a fluttery noise that jumps back and forth between channels, loss of separation, even fading of the signal entirely for a second or two. Quadrophonic phasing is nearly destroyed, which is why the signal may sound better in stereo than in quad.

The best solution is a different antenna position, or even an outside antenna. Directional FM antennas will discriminate against the unwanted signal. All-channel TV antennas are fine, too, since FM frequencies lie between channels 6 and 7, so you may be able to use your present TV setup with an antenna splitter from the local radio store. Cable TV often carries FM, and you may be able to hook your tuner up to that source.

The spec sheet may mention *de-emphasis*. This, and *pre-emphasis* are complementary treatments given the music by the FM station and your receiver. The station pre-emphasizes the signal—boosts its high frequencies—before transmission. When you receive the signal, the tuner de-emphasizes the highs an equal amount—and drops the hiss level appreciably as well. The U.S., Canadian and Mexican standard uses a resistor-capacitor combination with a *75-microsecond* time constant, which gives a great deal of high-frequency boost. Some European and Asian countries use less boost—a *50-microsecond* constant—and a few manufacturers offer a switch on the back so that you can adjust your radio for the country you're in.

So far we've been ignoring the AM broadcast band, and that isn't quite fair, even if a lot of manufacturers tend to do the same. AM radio is more susceptible to noise and interference, and has a somewhat limited frequency response, but with a little care you can hear some intriguing stuff—and with surprisingly good fidelity.

AM stations are spotted every ten kilohertz between 540 kHz and 1600 kHz. Since the signals they generate are twice as wide as the highest frequency in the music, this means they are limited to 5 khz as the top of their response. Higher frequencies would interfere with a station on an adjacent channel (if there *is* a station there . . .)

This leaves out three-quarters of the human hearing range, but only two musical octaves, and the least important two at that. Musical instruments only operate below these octaves—all that they contain are clicks and whines, harmonics from percussion and violins.

Besides this, AM signals in the daytime are limited in range, and interference from distant stations is less important. For this reason, some stations broadcast with responses up to eight or ten kHz—an extra octave—which is no worse than switching on the scratch filter on a dusty record.

Some manufacturers throw in AM more as an afterthought than as a feature. Some areas of the country do not *have* FM stations (try driving through western Kansas sometime); others are limited to a few stations playing Sinatra records. Listening to AM can be your only choice, and in any case it's a lot more interesting than you think.

Chapter Eleven

MILES OF MUSIC.

SECOND TO DISCS IN POPULARITY for storing music—and gaining rapidly—are the varied forms of magnetic tape. Tape is less convenient to use (why fast-forward through the whole reel when you could just lift the needle into the center of the disc?), but a great deal more flexible. You can record programs off the air, copy a disc, make live recordings, even multi-track like a recording studio, and with quality to rival the professionals if you're willing to put some bucks into it.

The bucks come in because tape recording is a complex process. Two persons using identical equipment to record identical programs can come up with completely different results in terms of quality. The difference is usually in the operator's knowledge of how the process works...so here's your edge. Magnetic tape itself is a sandwich. The bottom layer is a plastic ribbon called the *backing*; its purpose is simply to hold the rest of the sandwich in place. It may be anywhere from ¼ of a mil to 2 mils in thickness (a *mil* is a thousandth of an inch). We'll get back to that a little later.

The next layer of our sandwich is the *binder*, an adhesive which holds particles of magnetically-active material to the backing. These particles form the third layer of the tape, the layer which actually does the work of recording: The *oxide*. Tape oxide is made up of particles of iron oxides (FeO and Fe_2O_3), or sometimes of more exotic magnetic oxides (chromium dioxide, CrO_2, has enjoyed a recent vogue). These particles are magnetically charged in the recording process, and store that charge for later playback.

Putting a charge on a particle of magnetic material is

easy—just wave a magnet near it. But let's get a little more precise than that. Remember from your high school science classes that little experiment where you laid a piece of paper over a bar magnet, sprinkled iron filings over the paper and watched the filings line themselves up along the lines of magnetic force? We'll do the same; only instead of using a permanent magnet, we'll use an electromagnet.

Back in the 19th century a Danish physics professor named Hans Oersted discovered that a magnetic field was set up whenever you passed a current through a coil of wire (the discovering occurred right in class, which must have upset the principal). Wrap the wire around a bar of iron or steel, and the magnetic field was channeled through the bar—the bar became an electromagnet.

So far, perfect. Problem is, the magnetic field is scattered through space between the ends of the bar, too weak and too broad to be of much use to us. Let's bend the bar magnet into a horseshoe, so that the field is concentrated between the ends of the horseshoe.

Better, but still not very useful. If we bend the magnet into an almost complete circle, with the ends of the magnet just barely separated, we'll concentrate the field into the tiny gap. Finally, we'll file the ends of the magnet down into sharp wedges, and the field between the sharp ends of the magnet's poles will be tiny and intense, even with a very small current through our coil of wire. Almost all of the magnetic energy is focussed into a tiny oval of field between the two pole-points.

You didn't know it, but we just built a recording head. By passing our tape sandwich over the surface of the head, so that a tiny slice of it falls within the focussed field, we can control the magnetic charge on the oxide particles by controlling the amount of current through the coil, which changes the intensity of the magnetic field. The more current, the stronger the field and the greater the charge stored in the oxide.

This explains a few other things, too. If our tape passes near but not on the surface of the recording head (which is what happens when there's dirt on the head), it's lifted out of the focussed magnetic field, and gets little if any charge stored in the oxide particles. Suppose some oxide particles are scraped off a previous tape and lodge in the head gap; they conduct the magnetic *flux* (the lines of force) just like the bar of iron, and the field gets shorted out, never reaching the tape oxide. So your heads must be kept scrupulously clean, and the tape must pass over the heads *and touch them* to work at all.

Imagine a band of sandpaper running through your tape deck week after week. That's essentially what you have: Iron oxides are used in one type of sandpaper. Of course, this "sandpaper" is extremely fine, so it tends more to polish than abrade your tape heads, but if you look closely at the heads of a well-used deck (or a poorly-designed one, where the tape is not lifted off the heads during rewind and fast-forward) you'll see that the tape has worn a groove in the metal of the head. The better decks use all sorts of exotic, super-hard material to counteract this wearing-down; you pay for the better materials, too.

Have we left anything out? You bet. Oxide particles (and most magnetic materials) do not have a linear characteristic. That is, the magnetic charge is not proportional to the strength of the field. Worse, it exhibits a property called *hysteresis*, which makes for some tricky work in recording.

Tricky explaining, too: Let's take a single particle of oxide and put it in a magnetic field while we watch what happens. Increase the intensity of the field slowly from zero, and we find that the charge on the particle is...zero. Up to a certain point, the particle just lies there and picks up no charge at all (for reasons which are still not quite clear to the atomic physics boys, and way the hell beyond you and me...).

Beyond that point, the particle begins to pick up a charge, stronger as the field gets stronger. Our oxide particle is behaving itself...for a while. Once we reach a certain intensity, however, the particle calls it quits. It *saturates*, and no matter how intense we make the field it won't hold a charge stronger than the saturation level.

Okay, we know when we're licked. Let's back the field strength down again. But wait; our particle is holding onto its full charge. Drop the field to zero and it still holds the charge. Increase the field in the opposite direction (going from "positive" to "negative") and it *still* holds a positive charge...up to a certain point, where it begins to swing towards zero charge, and then into a negative charge, and finally into negative saturation.

Graph this rather odd behavior, field strength versus charge, and your graph will look like a fat, hollow, lopsided "S," which is our *hysteresis curve*. If it weren't for hysteresis, the particles would hold no charge and tape recording would not work at all; but we have to solve some problems with that fat "S" if we want hi-fi recording.

To get low-distortion recordings, we have to lay the field levels along the nearly-straight portions of the curve, which are the sides of the hollow "S," without hitting the saturation level. Keeping out of saturation is easy—just don't put too strong a signal into the head coils. Keeping the signal out of the hole in the center of the "S" requires some fast footwork.

The "fast footwork" is called *bias*. It's a signal at a frequency too high for humans to hear, mixed in with the audio signal we want to record. The bias signal swings through zero (and through the hole in the "S"), while the audio signal "rides on top" of the bias frequency and lands right in the straight portions of the "S" curve. That means distortion of the bias signal during playback, but we don't mind—it's too high a frequency for us to hear anyway. All we hear is the audio signal riding on top—and that was recorded in the linear section of the "S" curve, giving us the low-distortion storage we wanted.

Very neat—and very finicky, because that straight portion of the hysteresis curve is only a very small fraction of the whole curve. Worse yet, the curve is different for different formulas of oxide, for different thicknesses of oxide, for different particle sizes, even slightly different for different batches of the same brand and type of tape. (Now you're beginning to see why you pay $300 and up—and up—for a tape deck. A lot of engineering goes into one of these machines.)

Professional studios use one or two types of tape exclusively, and set up their recorders' bias and level adjustments for best results with those tape types. Most home recordists will put up with a little more distortion, and so the manufacturers adjust their machines at the factory for the middle of the range of curves for most brands of tapes. Recently, new types of high-energy and chromium-dioxide tapes, with very different curves, began appearing on the market, and manufacturers began putting switches on their decks so that the user could select the proper bias setting for each type. If you're a real purist, though, you'll have the service shop adjust the bias exactly for the type of tape you use, and then use that type exclusively.

Any other tricks? Yes: Random magnetic variations in the charge on the oxide particles come out as noise in the playback—the well-known "tape hiss" that sounds very much like white noise. To counteract the noise, tape recorders boost the high-frequency components of the audio signal in a process very similar to RIAA equalization on a disc. The equalization was standardized by broadcasters (who were the first to use tape recording professionally) in the 1950's; the NAB standard (named for the National Association of Broadcasters) is still used today. A standard treble boost is applied to the signal before recording, and a corresponding treble cut after playback.

In Europe there are two or three different equalization standards used in different countries. Some European decks have switches either on the front panel or inside the deck to select the desired EQ curve. Also, the old NAB standard was designed for use with the tapes of the 1950's and 1960's, and does not take best advantage of the high-energy and chromium tapes available today. Some recording studios have been experimenting with improved curves, notably John McKnight's Nagra-Master equalization.

As if that weren't enough, noise reduction techniques have been developed to both boost high-frequency components of an audio signal and compress their dynamic range. The best known of these is the Dolby system, which breaks the audio signal into a number of separate frequency bands, compresses each band individually, boosts the highs, then reverses the entire process on playback. The Dolby circuit has been superseded in professional use by improved devices such as the dbx system, but the low-cost Dolby "B" circuit, which works only on the high-frequency range of the signal, has taken the cassette market by storm. That's not surprising, since the cassette medium is only marginally "hi-fi" without it. To explain why, we first need to understand the mechanics of playback.

Run a current through a wire and you get a magnetic field—that's Oersted's discovery. An American named Joseph Henry found that it works both ways: Wave a magnet near a piece of wire, and you *induce* a current in the wire. Run a recorded tape across a tape head like our recording head, and a tiny current is generated in the coil, a current which can be amplified and used in our hi-fi. We can play back our tape.

The amount of current in the coil depends on the magnetic charge of the particles within the gap of the head's pole pieces—and that's where the trouble comes in. Let's say we want to record a single cycle of 10 kHz tone. At ten KiloHertz, which is ten thousand cycles per second, that single cycle only takes .0001 second. At seven and a half inches per second tape speed, the entire cycle only occupies .00075 inches of tape. If our head gap is more than .00075 inches across, the magnetic charge averages over the whole cycle, and cancels itself out—we get no playback.

Fortunately, head gaps are much smaller than that... but the smaller we make 'em, the more expensive they get. Worse, 7½ inches per second is a moderately high speed for home tape recording. Three and three-fourths i.p.s. is very common, and cassette decks are designed to work at 1-7/8 i.p.s. The slower your tape moves, the more your high-frequency response suffers, because a greater fraction of the cycle is crammed into the head gap and cancels itself out.

Problem number two: Random variations in the magnetic charge, which play back as noise. If you spread a cycle of your recorded signal over more tape—that is, if you record at higher speed—these variations tend to cancel out. Recording studios make their master tapes at 15 i.p.s. and even 30 i.p.s., for best noise and high-frequency response. At cassette-tape speeds, it's a real problem, and Dolby or some other type of noise reduction is a must for clean recording.

So much for the mechanics of tape recording. You can use tape in three different formats: Open-reel, cassette and cartridge. The first machines were open-reel designs; in fact, the very first were built in Germany during World War II, and used to broadcast Hitler's speeches over Nazi radio. They drove Allied intelligence crazy, because they knew Hitler wasn't in the studio but they couldn't figure out how his voice was.

After the war, they found out: Metal oxides on strips of paper tape, moved past the gap in a small electromagnet. In the best tradition of American enterprise, the intelligence officers who found the equipment took it home to the States and proceeded to found a company—Ampex—and an industry. Broadcasters began to use the new machines because they allowed shows to be assembled piece by piece, splicing the good takes into the tape and the bad takes out, which couldn't be done with disc recording. Recording studios soon followed suit, and the Great American Home-Entertainment Industry was off into high fidelity.

Cartridges were the next tape format to be invented. William Lear (the man who developed the executive jet, not the man who invented Archie Bunker) wound a reel of specially-lubricated tape, pulled one end out from the center hub and spliced it back to the outer end in an endless loop, then put the whole thing in a plastic case. Presto: A one-reel device that cycled the tape through the machine over and over again. Problem: You couldn't back it up, you had to wait for the whole tape to play through. Solution: Ignore the problem.

Cartridge tapes were snapped up by radio stations as a solution to the hassles of commercial announcements on disc. Special equipment would play the cartridge, let it cycle through until the beginning of the commercial was sensed, then stop it at that point ready to play again. The quality has since been improved, but the basic design is still the same, and it's used in almost every broadcast station in the country.

For home use—or more accurately, for use in automobiles—a similar design was developed. Just stuff the cartridge into your car deck and hit the button. By the way, the broadcast cartridge is slightly different from the standard eight-track—the pinch roller (which squeezes the tape against the spinning capstan to move the tape) is built into the cartridge in home carts, into the machine with broadcasters.

Unfortunately, you can only fit about ten minutes of tape into the standard cartridge—how do you put an entire forty-minute album on it? Simple. Just split the quarter-inch width of the tape into eight portions, each 1/32nd inch wide. Lay the first stereo program into the first and fifth *tracks*—each band is called a *track*—the next into #2 and #6, and so on. When you come to the end of the first program, a sensor picks up a piece of metal foil on the tape, shifts the playback head down slightly so that it's picking up tracks 2 and 6, and we're in business. At the next foil-sense, go to 3 & 7, then 4 & 8, then back to 1 & 5.

Eight tracks...oh, *that's* how it got the name! But aren't the tracks awfully narrow?

You bet. They're actually even narrower than 1/32", because there has to be a tiny *guard band* between each pair of tracks to keep one program from leaking over into the adjacent track. The narrow bands mean that you don't have many oxide particles carrying the program, so the noise doesn't average out too well, and the frequency response suffers. Eight-tracks are not known for their fidelity, just for convenience. But the technical limitations account for only half of that reputation.

In a car, the noise outside and in tends to drown out the worst of the cartridge's problems, so manufacturers out for a quick buck (and there are many of them...) cheapen their products as much as they can to get the price edge and the profit margin. Just as well—a huge fraction of the eight-track tapes on the market are not made by the legitimate manufacturers, but instead are pirated copies, often on poor-quality tape that flakes and sheds oxide, badly recorded and duplicated even worse. Why not? If they don't mind ripping off the artists, why should they worry about you? For these reasons, eight-track decks have never really gotten hold of the hi-fi market.

They are slowly losing their car market, too, to the cassette format. Developed in the early '60's by Phillips in the Netherlands, the cassette format offers the convenience of cartridge with added flexibility (you can fast-forward, rewind, and record your own on home equipment which was never very easy to find for eight-track) and quite acceptable fidelity. An early cassette designed by RCA never caught on, but the Phillips/Norelco cassette took off, first with people who needed cheap, portable recording facilities (for dictation, correspondence and so forth), then with the addition of Dolby in home high fidelity systems.

Cassettes use two or four tracks on a width of tape just slightly greater than 1/8th inch. For mono, the first signal is recorded along the top track, then the tape is flipped over and a second mono signal is recorded on the other track in the opposite direction. In stereo, program one goes on tracks 1 and 3 (numbered from top to bottom), then the cassette is flipped and tracks 2 and 4 get their program in the other direction. Stereo cassette tracks aren't much wider than eight-track bands, and the tape speed is a quarter of eight-track. Still, because the cassette package holds the tape more firmly and steadily in the machine, cassette fidelity and eight-track fidelity are theoretically pretty close...and practically, cassette decks are built to a much higher quality than most eight-track decks.

For sheer fidelity, however, neither eight-track nor cassette can touch a properly-designed open-reel machine. The advertisements will try to tell you that cassette decks can equal the performance of open-reel. Sorry. It's true that today's better cassette decks can satisfy all but the most finicky of ears, but match cassette against open-reel with equal conditions (Dolby, high-quality tape, normal tape speeds) and the open-reel will win every time. Besides, open-reel tape allows splicing and other manipulations that are inconvenient at best with cassette (and downright impossible with eight-track).

Open-reel machines of comparable quality are slightly

more expensive than cassette decks, because some of the machinery built into the cassette body (tape guides, tape spools, bearings) must be built into the open-reel deck, and with greater ruggedness since you can't throw it away when it wears out. The main difference, of course, is size—there's simply more metal in an open-reel deck. And open-reel is less popular (because less convenient) so it's more expensive to make each machine.

Open-reel uses one-, two-, and four-track formats. Professional mono machines use the full width of the tape for a single mono program; home mono machines use half-track, one program on each of two tracks, recorded in opposite directions. For stereo, the professional machine records two tracks, each half the width of the tape, in the same direction; home decks usually use quarter-track widths to allow flipping the tape and recording a second stereo program. Most home quad decks, and a very few professional machines, use four tracks in the same direction for four-channel; most of the pros go to wider tape for more channels: One-half inch tape for quad, 1" for eight tracks, 2" for sixteen. Two-inch is the widest tape generally available for recording; for more channels, recording studios will slave two machines together to act as if they were a single deck. (One designer, however, built a 40-track machine using 2" tape. Took him five years to unload the thing, too.)

Less expensive open-reel machines, and almost all cassette decks, use a single motor to spin the capstan, with a series of pulleys or idlers to transfer motion to the supply and take-up reels. These also usually have two tape heads, one for erasing a tape with a bias-like signal, one which is switched between recording and playback functions.

The more expensive open-reel decks (and the *most* expensive cassette decks) use *three* motors, one which simply spins the capstan, and *spooling motors* on takeup and supply reels. The extra motors help to smooth out variations in tape speed, and this cuts down on *wow and flutter*, which shows up as warbles and sirens in your recorded material. Unfortunately, three motors are more expensive than one.

Three heads are better than two in tape decks: The better decks will have separate record and playback heads, which allows you to listen to the material you're recording as it goes onto tape, but played back a split-second after it is recorded from the tape itself. The tape passes over the erase head first (blanking the tape), then over the record head (recording the program), then over the play head, allowing you to correct any problems that may arise before the program is over. There is another technical advantage to three heads, too: The optimum playback head gap is slightly smaller than the optimum recording gap, and the use of separate heads will improve frequency response and noise performance slightly. Every little bit helps.

Some tips for the home recordist: Keep that machine *clean!* Get some cotton swabs and a bottle of good head cleaner from your hi-fi store. If you can't find head cleaner, use isopropyl alcohol, available from your drug store. Do *not* use rubbing alcohol; it leaves a gummy film on the heads. Note also that "head cleaning swabs" and "Q-tips" or "surgical applicators" are identical, except you'll pay more for head-cleaning swabs. So don't.

Clean everything that the tape passes over: Heads, capstan, pinch roller, tape guides, flywheel. Your equipment manual will explain how to clean the machine. Read it.

Half-mil tape is fragile and bends or stretches easily; it also allows more of the magnetic charge on one layer of tape to leak through to the layer above and below (called "print-through"). One-mil tape is better for home use, and not that much more expensive per minute of playing time. One-and-a-half mil and thicker is standard for professional recording, so if you're really into fidelity, it's a good bet. Similarly, in cassettes the C-30 and C-60 tapes are thicker than the C-90's and give better print-through performance. C-120 is pretty useless for hi-fi, and should be reserved for class lectures on your portable.

Low-noise tapes, high-energy tapes and chromium tapes are capable of superior performance, but only if your machine is adjusted for them. If it is not adjusted, or doesn't have a switch for chromium or high-energy tapes, you might as well save your money and use the regular stuff.

Chapter Twelve

THE LOUDEST AIN'T NECESSARILY THE BEST.

Having picked up, preamplified, processed, modified, and amplified our musical signals, we will have to turn them back into sound before we can listen to them. For this we will need some high-linearity electroacoustic transducers.

Some *what*?

Some speakers.

It's sad but true that many hi-fi hunters let down their guard when buying speakers, and this is really too bad. Speakers have been called—quite accurately—the weakest link ih a hi-fi system. Modern circuitry has brought electronic noise and distortion to imperceptible levels; phono pickups and tape decks are not too far behind. But speaker design has lagged behind—there are so-called "high fidelity" speaker systems on the market today with distortion levels of five and ten per cent! And worse, people are buying them, convinced that they're getting true reproduction of their music.

Speaker design and marketing are probably the worst examples of hype in hi-fi. A friend of mine once remarked that "any idiot with a table saw and a soldering iron can go into manufacturing speakers—and a lot of idiots have." (It's only fair to mention that my friend has a table saw and a soldering iron, and manufactures speakers for a living . . .) The hi-fi magazines are full of fluff about stunning new advances in speaker design, and every year four or five manufacturers appear with "new" techniques that they guarantee will revolutionize the sound industry.

With that in mind, let's review a little history. The basic idea of a loudspeaker is to take electrical energy and move air around with it (unless you happen to be a fish, in which case we move water. Most of us aren't). The most common type of speaker used today—the most common type *ever* used—is the dynamic moving-coil speaker. It was invented in the 1920's, and with the addition of permanent magnets in place of battery-operated field coils, remains pretty much unchanged today. Electrostatic speakers (which somebody "introduces" at regular intervals of about two years) have been around since the 1940's. The physics of horn-type speakers (which the ads are presently trumpeting as the ultimate development in speaker technology) were mathematically set out in the early 1800's; and the system many experts acknowledge is the finest on the market today—the Klipschorn folded corner-horn system—was invented in 1942 by Paul Klipsch and is so old that the basic design patents have expired.

This is not to say that no one is coming up with anything new. Dr. Oskar Heil's tweeter and the Ohm transmission-line system show promise (and, alas, a few bugs yet); but by and large, the best available speakers are not sudden breakthroughs in design, but careful, well-considered improvements on the basic model.

Ah, yes, the basic model: The all-new 1924 Dynamic Loudspeaker. Battery-energised *field coils* provide a steady magnetic field; current through the *voice coil* (from the radio) provides a varying magnetic field in time with the music. Magnetic repulsion and attraction between the two fields moves the voice coil and the paper *cone* attached to it. A flexible *surround* at the cone edge, and a flexible *spider* hold the cone in position and spring it back in place when the magnetic fields are through pushing it around. A *basket* serves as framework to hold the whole thing together. And that's it.

The modern dynamic speaker replaces the field coil with a permanent magnet, removing the need for an extra battery. Aside from that one refinement, the dynamic speaker has remained pretty much unchanged. By the way, the name "dynamic" is used for two reasons: First, because it moves (and that's what "dynamic" means). Second, because the word is catchy—the advertising boys were on the job even way back then.

The other major speaker type is the *electrostatic*. In this type, a flexible membrane (the *diaphragm*) is suspended between two conductive screens, one charged to a high positive voltage and the other charged negative, both stationary. By varying the charge on the diaphragm in time with the music, the diaphragm is attracted first to one screen, then the other, pushing the air along with it. Electrostatic tweeters from time to time appear with great fanfare on the market, because they are capable of very clean high-frequency reproduction (they don't have those heavy metal voice coils to push around). Their disadvantages are the requirement for a power supply to charge the screens, difficulty of manufacture for big screens(which is why most electrostatics are tweeters), and the fact that sparks will fly on humid days, the first time Junior pokes a needle into the screen, or the first time Dad turns the volume way up on a castanet concerto. They're really pretty fragile little suckers.

There is a third type of speaker: Dr. Heil's "Air Motion Transformer." In this device, flat "wire" is bonded to a thin plastic sheet, which is then accordion-folded between the poles of a magnet. Current from the amplifier running through these flat wires reacts with the magnetic field and squeezes or opens the folds of the plastic (depending on the direction of current flow), and squeezes or sucks air in and out of each fold. The first amt's were tweeters, but Heil and his licensee ESS are reported to be working on a woofer version. Stay tuned for the latest details.

The advertisements will tell you that horns are the very best kinds of speakers there are. Shows what the ad people know: A horn isn't a *speaker* type at all, but rather a kind of enclosure. This brings us to another important subject in the speaker world: The kind of box you put it in.

First let's look at a speaker out of its box, naked to the world. As the speaker cone vibrates back and forth, sending out sound waves, we see that sound waves radiate from the *back* of the speaker cone as well as from the front, and furthermore, when the front of the cone is pushing air, the back of the cone is sucking air in. If a speaker is left out of its box, these two sound waves tend to cancel—the air just jumps back and forth around the edge of the speaker, and you don't hear anything (actually, the waves don't cancel *completely*; but the lower the frequency of the sound, the more time the air has to get around the edge, and the more complete the cancellation. In terms of sound, this means that the bass response drops to just about zero, as far as your ears are concerned, even though the cone may be moving back and forth like a mad thing).

One solution is to put the speaker in a closed box, stuffed with cotton or some other absorbent material to soak up the

waves off the back of the cone. This technique is called an *infinite baffle* because the back wave is "baffled" from finding a way to the listener's ear. (Some people claim "infinite baffle" is what the ad agency is hired to do.)

Back in the 1930's, when the movie theaters were all converting to talkies, you *couldn't* always get a bigger amplifier—with the vacuum tubes they had then, a ten-watt amplifier was a bruiser, the equivalent of today's 250-watt monsters. They couldn't afford to just "throw away" half those watts into a box; instead, they found a way to use them.

By putting the speaker into a box with a second hole cut in it, and a relatively long tube attached to the hole, they found a use for those back waves. By the time the back wave traveled the length of the tube, the front of the speaker cone had finished pushing the air and started pulling back again. Now the suction wave from the back of the speaker reinforced the front wave, rather than cancelling. Thus the *bass reflex* cabinet was born (so-called because the bass was "reflected" inside the box).

The problem with the bass-reflex was that the tube length and the frequency of the sound wave were critically related. Some waves reinforced; others didn't change direction on the front wave fast enough; still others changed *so* fast that they went past reinforcing and started to cancel again. The frequency response curve of the bass reflex looked like a roller coaster (and still does). Engineers built mazes, longer and shorter tubes, labyrinths, tapering tubes, blocks and baffles, absorbent material and weirdly-shaped holes into the box to try to flatten out the response, with varying degrees of success. The bass reflex got more sound for less power, but at the cost of screwing up a flat frequency response curve.

In the early 1960's, an Australian engineer named A.J. Thiele took a good hard look at the bass reflex design with the help of a computer and came up with a number of radical changes in the system. Of course, everyone knows that Australians drink a lot of beer and raise kangaroos, and that's about it; so American designers didn't pay much attention to Thiele's findings until just recently. The few people who have experimented with Thiele's results have found that he knew just exactly what he was talking about... and you may be seeing the comeback of the bass reflex design. Just remember, though, when you read an ad about an "exciting new tuned-port bass reflex loudspeaker"—that "exciting new" speaker has been around longer than *you* have.

The standard infinite baffle system has a smoother frequency response than the bass reflex, but there's a catch—for a decent bass with a 1950's vintage speaker, the cabinet had to be 10 or 20 cubic feet in volume—a trifle hard to camouflage with a vase full of flowers. The key to a solution lay in the stiff suspension (spider and surround) of the speaker. A New York University acoustics professor named Edgar Villchur reasoned that a loose, highly flexible suspension would let the speaker cone travel freely and the long cone travel would move enough air for good bass. To restore the cone to its resting position all you would have to do is to mount the loosely-suspended speaker in an airtight box and let the air inside do the work, pushing or pulling the cone into place until the pressures inside and outside the box were equal. Villchur proceeded to do just that, and in so doing stood the speaker industry on its ear. Ironically, the patent office decided that Villchur's innovation was so simple it didn't deserve a patent, so Edgar had to settle for copyrighting the name "acoustic suspension" and founding Acoustic Research, Inc. The AR speaker became the most famous system of the hi-fi boom, and today four out of every five speakers sold are acoustic suspension types.

Suppose, for a minute, that we put our speaker at the end of a long tube. Suppose too that we flare the tube gradually, so that the sound doesn't rattle around inside like a marble in a barrel, but instead spreads out into the air without reflection. Presto: The exponential horn (which is even older—a mathematician named Helmholtz and his drinking buddies worked out the math of the exponential horn back in the 1830's).

The "regular" speaker works by brute force, shoving the air back and forth by sheer motion. The majority of the air molecules (being lazy like the rest of us), just slide off to the side and don't make any sound at all. In the horn speaker, the air is constricted by the throat of the horn—it *can't* fall off to the side, it can't just sit there, it *has* to carry the sound wave. This makes the horn speaker much more efficient than the bass reflex or the infinite baffle—and because it's so much more efficient, you don't have to push it as hard to get the same amount of sound. Not only does this mean a smaller amplifier, but also the diaphragm of the driver doesn't move as far, and so you don't get nonlinearities—distortion—from inertia and the friction of the surround. This is the reason horns sound so good; they're just not working very hard.

Incidentally, the "driver" is just our old friend the dynamic speaker (some horns have used electrostatic or even more exotic drivers as well), with the cone replaced by a spherical dome, which is a stronger structure that won't bend under the pressures built up at the end of the horn. Those pressures can be quite large—a couple of pounds per square inch or more. All the horn does is to sit there and corral the air molecules into the right place.

The shape of the horn is quite critical— too long and narrow, and it will sound hollow. Too wide, and it won't do any good. Any roughnesses or irregularities, and the high frequencies will bounce off the rough spots and cause harshness and distortion. And the size of the horn depends on the frequency of the notes it has to play. Highs may only require a horn a few inches long and wide; a bass note at 20 Hz. needs a horn 32 feet long.

This can pose a problem for somebody who wants a horn speaker in his den (one fanatic I knew built a horn through the wall and out into his garage; he drove the Cadillac under the throat every night). Fortunately, a gentleman named Paul Klipsch found that you could fold a horn around itself without severely damaging its properties. This takes advantage of a peculiar property of waves: They tend to ignore corners and holes that they can't fit into. You've noticed the same effect with radio waves: If you drive through a tunnel, the AM stations (with waves half a mile long) fade out. The FM stations (with waves a few feet long) keep right on playing. Similarly, the bass notes ignore the corners, bends, and curves in a folded horn. Of course, the higher notes *won't* ignore the corners, but by that time the waves are small enough to play through another horn, this one only a few feet long.

That brings us neatly to another subject: Crossovers. Play a bass note through a 12" woofer and the cone moves back and forth two or three hundred times a second without working up a sweat. But play a *high* note, that tries to shake the cone fifteen or twenty *thousand* times a second, and the paper cone will bend and flex and otherwise act very silly, if it doesn't rip itself right off the voice coil. You need a Papa Bear speaker for the bass notes, a Mama Bear speaker for the middle range, and a Baby Bear speaker or two for the itty-bitty high notes. Some less costly speakers try to get away without Mama, since housewives are too extravagant, but as everyone knows, Dad and Junior alone don't get the housework done as well by themselves. (Translation:

Making the Connection

Three-way speakers are usually better than two-way speakers.) You also need some sort of circuit to separate out the lows, midranges, and highs and send them to the proper speaker. This is called a *crossover network*, because as the frequency of the sound goes past the design frequency, the sound crosses over from one speaker to the other. Actually, crossover networks are nothing more than glorified tone control networks—the woofer network "turns down the treble" (in its case everything over a couple hundred hertz); the tweeter's "turns down" the bass (everything below five thousand hertz); the midrange blocks out the tweeter range and the woofer range and passes everything in between.

There are crossovers and crossovers. There are superfancy electronic crossovers that separate out the different frequencies and feed them to separate amps for each speaker, so that the electrical energy won't be wasted in the crossover's slight resistances. At the other end of the scale, some of the cheapest speakers only put a crossover on the tweeter, on the theory that the woofer will just give up on the high frequencies. In fact, the woofer *doesn't* give up on the highs—it tries and fails, but in trying it buzzes and rattles and otherwise goes *boing* when it should have gone *mmmmmmm*. You get what you pay for.

Designing crossovers gets into some moderately hairy math, but the nice thing about it is that there are lots of books by people who have already done the math at your local library. Even better, Speakerlab, Inc. (whose address you'll find listed at the back of the book) put out a Crossover Design Guide for a quarter that'll tell you just exactly what to put in your crossover and how. All you have to do is look at the charts.

So what's so hard about designing speaker systems? Nothing. But the speaker companies aren't going to go out of their way to tell you that you're as smart as they are. They'd rather have you a little in awe of them. Honestly, the hardest part of making a speaker system is to get the box looking good—getting it to *sound* good is a certified snap. There are many fine systems on the market, and if you're concerned with the brand name or don't want to bother with the table saw, by all means buy one; the proof is in the listening.

But if you think you might want to try building a system yourself, go ahead: You'll surprise yourself very pleasantly. End of commercial.

INTERLUDE:

By this time you've probably been to visit all the hi-fi stores in Your Town.

Sometimes it's a little difficult to track down those who are Over The Hills and Far Away.

For those of you who want to save by rolling your own, we present a short and non-exhaustive list of **Kit Makers:**

Dynaco, Inc. Box 88, Blackwood, N.J. 08012

Excellent tuner, preamp, power amp, integrated amp kits, with idiotproof assembly instructions.

Heath Company, Benton Harbor, Mich. 49022

Grandaddy of the kitmakers, and inventor of the idiotproof assembly manual. Tends towards kits-as-hobby instead of kits-as-savings, but pretty good hi-fi stuff.

Speakerlab, Inc. 5500 35th Ave. NE, Seattle 98105

Excellent speaker kits in all ranges, as well as raw speaker components, design info and general good karma.

Southwest Technical Products Corporation, 219 W. Rhapsody, San Antonio, TX 78216

Clean and simple hi-fi component kits and assorted widgets for the electron freaks. Note: their instructions are **not** idiotproof.

PAIA Electronics, Box A14359, Oklahoma City, OK 73114

Synthesizer kits and assorted black boxes. Neat stuff once you already own a hi-fi.

Apologies to the many other kitmakers whose addresses I've lost or whose acquaintances I have yet to make.

Chapter Thirteen

OLD WIVES' TALES.

Over the years any number of legends and myths have grown up around stereo equipment, balloons that deserve to have a few holes punched in them. In this chapter we'll examine a few of them and see just how much truth there is to each.

Old Wives' Tale #1: "The only proper way to listen to high fidelity sound is with no equalisation, the tone controls set dead center."

This notion was dogma with the early purists. When hi-fi first came on the scene no one had actually heard reproduced sound that could compete with the actual event, and the purists maintained that the human ear was a poor instrument for determining tonal balance (never mind that this same poor human ear was the device they were planning to use to hear delicate nuances of music once the tone controls were set flat). Instead they had laboratory tests and charts and specifications which "proved" that their amplifiers reproduced audio with no discernible alteration of the frequency balance, which was the only way you could insure truly accurate reproduction of the sound—sound with a high fidelity to the original (which is, of course, how the name came about).

What this theory *didn't* take into account was the frequency variations of the speakers, of the phonograph cartridge, room acoustics, and the fact that the recording engineer probably did some diddling with the frequency spectrum before the music ever got on record, all of which tended to obviate the super-flatness of the amplifier.

The tone controls are there for a reason—to make up for the

effect of the room, the speakers, and everything else. There is nothing heretical about using them to make the music sound more enjoyable—after all, the whole point is to entertain *you*, right? However, don't fall into the trap of

Old Wives' Tale #2: "The best way to listen to hi-fi is with the bass jacked all the way up."

This is a common mistake among people who have just started getting into hi-fi. It's a natural mistake, too: Chances are pretty good that they haven't *heard* real sound before, only tinny little tunes out of a table radio. The first difference they notice is that, all of a sudden, there's all that bass they never heard before, and, if some bass is good, a lot of bass must be great, right?

Wrong. Actually, listening to high fidelity sound is a learning experience, and most people pass through three stages. At first they are overwhelmed by the bass, and turn the control all the way up. After they get used to that (or tired of it . .), they begin to notice the treble—the high, sharp harmonics of the violins and cymbals that weren't there before either, but weren't as obvious as the bass. This phase tends to pass fairly quickly, both because the listener has learned the failures of excess in the bass phase, and because whatever background hiss and record noise is present usually shows up in the treble.

In the third phase, the listener has grown out of tone-twiddling and has started listening to the music. It's at this point that he begins to learn the difference between distorted and undistorted reproduction, and this is where it gets expensive. The only way to cure distortion *in your amplifier* is to sell it to somebody else and get a better one. Now you're on your way to The Search For The Perfect Sound (and you've started your neighbor out in high fidelity).

Hi-fi enthusiasm is something like venereal disease—it's an epidemic you spread to your best friends.

Old Wives' Tale #3; "Rock music will make you sterile"

This is why your parents don't want you to listen to rock; they want grandchildren to dandle on their knees.

Some researchers experimented with background music around animals. They found that classical music stimulated cows to produce more milk, while rock music slowed production and the milk that was produced tended to go sour very easily. Rats and guinea pigs exposed to high levels of rock stopped reproducing, and, in extreme cases, curled up in a corner and refused to eat.

To date no one has done similar studies on teenagers, but it seems unlikely that a group that survives on a steady diet of pizza and McDonald's hamburgers can be damaged by music of any kind.

Old Wives' Tale #4: "Rock music will make you deaf."

There's some truth to this one. Scientists have measured sound levels at rock concerts as much as ten times *louder* than the noise directly under a 747 taking off. The real point is, loud noises of *any* kind will make you deaf, if you listen to them often enough. Many professional rock musicians have been found to have moderate to severe—and permanent—ear damage from standing in front of banks of guitar amps.

Most home hi-fi setups simply do not have the capability to produce that kind of dangerous sound level, and even if they did, the family in the apartment down the hall would shut you down long before you did yourself any real damage.

A few tips to tell if you're hurting yourself:

A "gut-shaking bass" is really just that; you "hear" bass more with your stomach and viscera than with your ears. Treble, on the other hand, is quite literally ear-piercing—turning it all the way up with a screaming guitar can damage your ears. (It will also probably blow out your tweeter..)

Headphones can be especially dangerous, because you can turn them up to dangerous levels without disturbing the rest of the household. After using headphones for a half-hour or so, take them off and talk to someone in the room. If you have trouble hearing him, they were too loud. Leave them off and go have a pizza.

This sort of temporary hearing loss is called "threshold shift"—the muscles and nerves in your ears are fatigued and not as responsive as they were earlier in the day. The louder the sound you're listening to, the quicker your ears will become fatigued. Recording engineers set their monitor levels to a point where they'll be able to listen for six or eight hours before their ears go out—and then they go home and rest.

This is very important. A threshold shift is usually temporary, and doesn't do irreparable damage to your ears, *if you rest them afterwards.* Without rest, the temporary loss becomes permanent—not suddenly, but gradually and without warning, until you suddenly realize you can't understand a word of what the guy across the room is saying. If you find you're having trouble hearing, or your ears are ringing (a sign of *severe* threshold shift), go somewhere quiet and take a nap or read a book. It's really worth it.

Hearing loss of this type is almost inevitable in a noisy society like ours—researchers have found many forty-year-olds who can't hear a thing above ten kilohertz. It's the treble range that goes first, and this is the frequency band that gives definition to consonants in speech. This is the reason why older persons often complain that "young people all mumble today"—they can't hear the sharp clicks and sibilants in speech, but only a kind of slurred low rumble. High-frequency noises—typewriters, radio, traffic, whistles, etc.—tend to accelerate this hearing loss. Plan your listening accordingly.

Old Wives' Tale #5: "Rock music needs more power than classical."

On the average this is true; the usual rock number will require more power from your amplifier for realistic reproduction than the average classical piece; but there are so many exceptions that it doesn't pay to buy low-power equipment for classical music.

What makes rock hungrier for watts is its preponderant bass, which carries the most acoustic power. Your average symphony orchestra has little demand for electric bass or synthesizer. However, anyone who has ever heard a good recording of the 1812 Overture (subtitled "concerto for cannonnade and church bells") knows that there's plenty of bass in the classical repertoire, too. Furthermore, rock music generally has very little dynamic range—it's all the same volume from one verse to the next. Classical, however, has a great dynamic range, which means the loud parts are *LOUD*... and need that much more power.

Old Wives' Tale #6: "FM is an inherently higher-fidelity medium than AM."

Not really. "AM" and "FM" merely refer to different ways to put information on a radio wave, not how much or what kind of information. What limits AM performance is history; the Federal Government, when it set up rules for AM broadcasting early in the 1920's, figured that a 5 khz. top frequency was all you really needed for good communication. Now, the physics of radio being what it is, a radio station with a top modulation frequency of 5 khz. uses 5 khz. of bandwidth on *either* side of their center frequency—ten kilohertz total. So the Federal Communications Commission (actually the Federal Radio Commission in those days) put stations every 10 khz. along the band.

Unfortunately, by the time people began to realize that there was music worth listening to above 5 khz., there just wasn't room to spread all the stations out a little, and the top modulation frequency remained limited to 5 khz.

A couple of other problems cropped up with AM, too. Spacing a station every ten khz. meant that, when two stations were close enough to your receiver for both to be heard, there was a constant 10 khz. whistle behind the music, from the interference between the two stations' carrier signals. The FCC tried to alleviate this by requiring stations on adjacent frequency assignments to be located at a great distance from each other (depending on how powerful each was), but late at night when the signals travel farther by bouncing off the ionosphere, the whistle is still there—and very annoying.

Another problem is that most electrical noise—lightning, ignition noise from cars, industrial noise—is amplitude modulated. This means the receiver can't tell the difference between a lightning stroke and a radio station...and so your radio gives you both.

The FM band, however, is located on a frequency range which does not travel much farther than line-of-sight, with plenty of room for everyone to stretch out. The FCC engineers limited FM broadcasters to a 15 khz. top frequency of modulation, which seems shortsighted today but in those days appeared to be a ridiculous amount of overkill. Furthermore, an FM receiver *could* tell the difference between an FM station and a lightning bolt, and all you heard was the station. These advantages caused the early hi-fi afficiandos to hail FM as a quality medium and give up on AM forever.

Don't forget about AM entirely, though. During the day, when signals only carry about 50 to 100 miles, there is little interference from distant stations, and stations can often get away with broadcasting frequencies as high as eight or ten khz.—no worse than when you cut in the scratch filter on the turntable. In good weather and under these conditions, AM classical stations often broadcast quite good quality music (technically speaking, at least), and even at night reception can be surprisingly good.

Old Wives' Tale #7: "But FM has stereo, and you can't broadcast stereo on AM."

At present, no, they are not doing stereo AM. But it has been done, on an experimental basis, by a couple of stations in Mexico. You'll remember we said AM stations use chunks of spectrum above and below their frequency? These are called *sidebands*, and what the experimenters did was to use each sideband independently, the upper sideband for the right channel and the lower sideband for the left. A special receiver—or two regular AM radios, one tuned slightly below and the other slightly above the station's carrier frequency—gave you quite respectable stereo reception.

At present the FCC has not given any stations in the U.S. permission to use this system, but some stations (afraid of being left out in the cold by FM Quad) have been petitioning to use it, and the day may yet come...

Old Wives' Tale #8: "Radio stations have much better equipment and fidelity than home sets."

Not by a long shot. Since the hi-fi boom first took off, in fact, the quality of consumer equipment has improved to the point where it is now audibly better than the equipment at most radio stations, and approaches the quality of good recording studios (which really *are* as good as technology permits).

There are a couple of reasons for this. First, radio station equipment is used 24 hours a day, day in and day out, by hamhanded jocks who bang it about and are as likely as not to throw it bodily on the floor if they get mad at something. It is usually serviced by an underpaid, overworked engineer who scarcely has time to repair it when it breaks, let alone practice preventive maintenance on it. So it has to be rugged and durable, and this usually means sacrificing a little performance in favor of sturdy construction. There *is* sturdy, durable, quality broadcast equipment, but it's usually expensive, which brings us to our next point.

Most station owners regard their station as a money machine, to be drained of the last penny they can wring out of it (sad, but true). A few enlightened managers (usually the exceptions to the above rule) buy the expensive stuff, knowing they'll more than make up the cost in increased audience and lower repair bills; the majority buy the least expensive equipment they can get away with and complain bitterly to the engineer when it breaks down. It's not uncommon to see a station operating with turntables and amplifiers 10 or 20 years old, held together for "just one more rating period" with masking tape and fervent prayers.

Old Wives' Tale #9: "But doesn't the FCC set requirements on a station's technical operation?"

The FCC requires a broadcast station (AM or FM) to be within a certain tolerance of their assigned frequency. They require that the "main microphone channel"—the microphone through which the announcer does most of the talking—meet certain requirements about noise, frequency response, and distortion. These required "standards," incidentally, can easily by surpassed by the average $100 receiver today—they haven't been materially changed in 30 years.

Aside from that, the FCC Rules and Regulations make no mention whatever of turntables, tape equipment, mixing consoles, or even any other microphone in the station. They could be drowning in noise and hum and the FCC wouldn't make a fuss.

To be sure, there is some incentive among stations (especially on FM) to broadcast a clean signal—too many listeners with good equipment and discerning ears would complain. And there *are* some stations and engineers that pride themselves on a sharp, clear signal. But all too often the signal is no better than it absolutely has to be. The manager may spend thousands of dollars on equipment to insure his signal is the *loudest* on the band, but the *cleanest* ...?

Uh-uh.

Chapter Fourteen

CARE AND FEEDING.

The stereo salesmar's job ends the minute you take your system out the door...but your job is just beginning. Setting the system up properly can make the difference between years of carefree operation and bimonthly visits to the repair shop.

Electronic equipment likes the same sorts of places that people do: Cool, clean and dry. Electronic components generate a great deal of heat—even the "passive" components like resistors and capacitors—heat that can literally cook the circuit into malfunction. Humidity is another enemy, as most components will be partially or totally shorted out by moisture. Add a dollop of dust to trap the heat and humidity inside and you have a dandy oven for baking the circuitry—which is fine for Colonel Sanders, but not for high fidelity equipment.

Properly-designed equipment is ventilated to carry away the excess heat, and with it the humidity. The ventilation usually depends on convection—hot air rising through the bottom of the set and out the top. Most manufacturers include in their instruction books a warning not to build the unit into a wall console, or otherwise impede the flow of air around it. I know a lot of people who have ignored this good advice—most of them I met when they asked me to fix their stereos.

First, set your stereo up across the room from where you're usually sitting—couches and chairs collect dust and puff out a cloud of it when you sit down. Second, do like your mother taught you and vacuum once in a while. Sweeping generates a lot of airborne dust, so avoid it if you can, but sweeping is better than nothing if your Hoover is on the fritz.

Keep the system out of the sun—not only does the sun

generate heat inside the set, but it can get the plastic dial hot enough to warp and bend, not to mention what it'll do to your precious records. You'll usually want to keep your records and tapes near the turntable, deck and amp, so plan a shady, cool spot somewhere in the room for everything.

Houseplants can be a problem, since they make a room more humid. A couple of plants in the windowsill won't hurt anybody, but an indoor greenhouse and a hi-fi system aren't compatible. (one lady I knew put her Boston fern *on top* of the hi-fi—the daily misting shorted out the amp and the heat of the system burned out the fern.)

Pick out a few hardy plants if you like and leave them in the room—throw everybody else out. This isn't a bad idea from the plants' point of view, either, since research has shown most houseplants will wilt when exposed to a lot of loud rock 'n' roll. (Fact!)

Take yourself into account, too; remember that you'll be adjusting volume and tone controls, tuning, watching the meters, loading records, putting the tone arm down, etcetera. It's a good idea to arrange things about chest height. Waist level is okay too, though you'll have to stoop to reach the tuning meter. It's hard to handle a tone arm gently from much below your belt buckle or much above your eyes, and your record collection may suffer if you put the turntable outside that range.

A good idea is to put the turntable *below* the receiver, on the second shelf, where it's easy to handle, and leave the controls on the amp up high enough to see as well as adjust. You may have to build a special shelf for all of this, but that's okay because it lets you make room for the records and tapes too.

Ah, yes: Records. Made from polyvinyl chloride, a petroleum derivative, which is quite hard over the short run, but essentially a fluid over a long period of time. What's that mean to you? It means file them vertically, with enough "breathing room" between them that they don't press against each other, but not so loosely that they lean at an angle. Stacking them horizontally tends to press the vinyl back into the grooves of the bottom few records, while leaning records will develop a dish-shaped warp that lifts the tone arm up and down. For some perverse reason, warped records will never warp "back" if you lean them the other way. They just get more complicated warps. And never, *never* leave one edge of a record lying over the end of a table unsupported!

Tapes should also be stored vertically, inside boxes to keep the dust out. Otherwise, gravity will tend to warp the plastic base of the tape and increase the wow and flutter of the playback (or make it impossible to record properly). Heat and humidity accelerate these warping processes, which is why we keep the records and tapes in the same cool, dry place we keep the system.

Now we come to the problem with this neat, compact package: You have to put the speakers somewhere else. Stacking the speakers on the same cinder-block-and-plank shelves you just put together is asking for trouble, and the trouble is called *acoustic feedback*. The tone arm/turntable combinations can pick up vibrations from the room—someone walking across the room, the couple fighting in the apartment upstairs, or the "vibrations" that are the sound coming from the speakers.

Acoustic feedback is caused by the speaker sound vibrating the stylus and sending a new signal to the amp, which sends it to the speakers, which shakes the stylus, which goes to the amp, which sends it to the speakers... It may sound like a low rumble or a howling moan, depending on how bad the problem is.

If you hear something but you're not sure it's acoustic feedback, turn down the volume. If it goes away, it *was* acoustic feedback. Turn it down; but this usually means listening at a lower volume than you had in mind. Don't despair, there's other ways.

In most homes, the floor and walls aren't "connected" very strongly to each other in an acoustic sense. By mounting your speakers on the wall and the system shelves on a floor-standing arrangement, feedback is often reduced quite drastically. Another trick, if you use floor-standing speakers, is to mount wall shelves.

If you don't want to rearrange the decor of your living room just to get good sound, try some foam rubber. The obvious place to put the foam is under the turntable, to isolate it from the vibrations in floor and wall. That's the obvious place, but it's not the *best* place. Try instead a layer of foam (or corrugated cardboard, or polystyrene foam, or even crumpled newspaper) under the *speakers*.

Putting your acoustic isolation under the speakers keeps the vibrations from being transmitted anywhere in the first place, except through the air where they're supposed to go. It also has the advantage of quieting the sound transfer through to the next apartment, a fact I'm sure your neighbors will enjoy.

It's always possible to get acoustic feedback if you turn the volume up high enough, even if the feedback is being picked up through the air alone. But you can minimize the problem simply by moving things around until you find a position that works cleanly. Wall-floor mounting and simple physical separation of speakers and phono is a good way to start, but often it's just a tedious matter of trial and error.

Speaker placement can be a problem... but proper speaker placement can also *solve* a problem, or at least help to alleviate it. By placing your speakers at a corner, where wall and floor or wall and ceiling meet, the bass response is improved. This is because the midrange and treble frequencies travel straight out in front of the speaker, while the bass frequencies are relatively non-directional and radiate in all directions from the speaker.

By placing the speaker against a wall, the bass which radiates out the back is reflected towards the listener. The effect is stronger if the speaker is placed where wall and floor (or ceiling) meet, and stronger still where floor or ceiling meet *two* walls, in the corner. The effective "extra" bass below 200 hz. or so can help the sound of an inexpensive speaker. Conversely, if there's too *much* low bass and you can't properly adjust it with the tone controls, try moving the speakers out of the corner or away from the wall.

This technique is useful for acoustic suspension speakers and smaller bass reflex designs. The larger bass reflex speakers usually do not improve markedly with this trick, while horn-type speakers have their own built-in "walls" (at least one horn-type speaker, however—the Klipschorn folded horn—is *designed* for use against a wall. The Klipschorn design uses a corner formed by two walls as the last section of the folded horn, and actually does not work properly unless it is properly placed in a corner).

If components are finicky things, records and tapes are downright hothouse lilies. We've already talked a bit about what warps records; let's look at a few other problems.

Radio stations will sometimes fire a disc jockey who is careless in handling records—it's literally easier for them to replace the jock than it is to replace that old out-of-print LP. Over the years broadcasters have developed a number of tricks for the care and maintenance of records.

Trick #1: Take the plastic wrapper off the outside of the record jacket and throw it away. If it gets heated (or cooled), it can shrink and warp the record inside. Trick number 2 usually raises some eyebrows: Take the paper liner inside the jacket off the record and throw *it* away.

This advice comes, of all people, from the Mattell Toy company. Back when Chatty Cathy was starting to yap, the Mattell researchers wanted to maximize the life of the little discs that made Cathy mouth off. They found that the paper liners inside LP's tended to collect dirt from the record as it was put in and taken out many times, dirt that was then ground into the grooves of the record. Leaving an empty jacket that could be squeezed open for the dirt to fall out improved the life of the record. Plastic jacket liners that were supposed to be less abrasive than paper liners actually attracted *more* dirt through static electricity.

Trick #3 is an extension of the basic rule of all record-handling: *Never* touch the grooves of the record, with *anything!* So when you take a record out of its jacket, use trick #3: Squeeze the record jacket open between your chest and your left hand, slipping the disc out supported by the edges in the sides of the jacket. As it slides out, catch the edge at the base of your right-hand thumb, and slip your right-hand fingers under the record label. You can now support the record in one hand while you put the jacket down.

Supporting the open record between the palms of your hands is easy now, and you'll find you can flip the record without difficulty and without touching the grooves.

Replacing the record in the jacket is just as easy. Pick it up with your palms (be sure you don't reach under the record and touch the grooves with your fingertips), transfer to one hand with thumb on edge and fingers on label, squeeze open the jacket between chest and left hand, and let the record slide smoothly into the jacket. Now put it back in the shelf and you can get another record.

What? You already have another record? WRONG!! Put one record away before you've gotten the next, or at least make sure it's in its proper jacket. Otherwise you're too likely to pile one record on top of another, which scrapes dust between both; or you'll leave it out and it will warp; or you'll just get them in the wrong jacket and really confuse yourself next time you want to hear that LP.

There are a number of devices on the market for record care, some of which are worse than the problems they're supposed to solve. Prime among these are those chemically-treated dustrags which sell for a buck or two. They're fine for about the first five wipes or so, after which they're dirtier than the records you're cleaning with them. Some of them also leave behind a gummy residue that is worse than the static it's supposed to suppress in keeping dirt in the grooves.

The handiest device for maintaining records in good condition is Cecil Watts' *Dust Bug*, or one of the many imitators. This consists of a second "tone arm," at the end of which you'll find a round plush pad and a brush with long, sharp bristles. The bristles dig down into the groove and drag out dust particles—being softer than the vinyl, they do no harm to the record. The plush pad then collects these particles as the record turns, and at the ends of the record they can be flicked away with a finger or with the bottle of anti-static fluid that comes with the Dust Bug. The fluid is best, since you get a little skin oil on the pad when you give it the finger; if you've run out of fluid, distilled water, or ordinary (soft) tap water will do almost as well.

For really dirty records, the Discwasher system or Cecil Watts' Disc Preener are good ideas—these are velvet pads with anti-static fluid inside. The short bristles of the velvet dig the dirt out of the grooves as the record turns—just hold the device gently against the record as it turns once around.

Preeners don't clean quite as deeply as Dust Bugs, and it's a good idea to use both each time you play a record—Preener before, Dust Bug during, and Preener again after play, to pick up any dust that may have settled out of the air while the record was playing. Your friends may look at you weird, but your records will sound good for a lot longer than theirs will.

As a last desperate measure for an incredibly dirty record, you can take it into the shower with you. Really! Just use lukewarm water at low pressure and a mild liquid soap or detergent—dishwashing soap is fine. Once you've got the suds rinsed off, let it drip-dry in a clean, non-dusty room—and don't try to hurry the drying with your sister's hair-dryer, or you'll be worse off than when you started.

Tape is a different matter—it doesn't get dirty as much as records do, but when it does it is much harder to clean off. The cure here is mainly to keep the *machine* clean, and let the tape take care of itself.

If by some chance you've accidentally spilled tape on the

floor, you should just throw it away. The dirt and grit it has picked up will damage your deck as if you'd thrown a handful of sand into the mechanism. If you *have* to save the tape, make sure you clean it with a dry cotton rag, foot by foot, before you play the tape—and keep the rag clean as you do it, too.

Keeping the deck clean is simple but tedious—get a bunch of cotton swabs and some head cleaning solution, and vigorously rub every place the tape passes over, heads, guides, trip switches and rollers. Rubbing alcohol is a fair substitute if you can't get proper head cleaner, though it leaves a residue that you have to clean off with a dry swab later. Head cleaner has a lubricant built in that helps keep the tape flowing cleanly, so use it if at all possible. Do this once every ten hours or so of use—more often in daily recording—or plan on trading the deck in two or three years early.

Storing tape is fairly easy—just put it in that old cool, clean, dry place, stored on reels or cassettes or cartridges, preferably inside a box to keep out dust.

The best way to store open-reel or cassette tapes is "tails out." Which means backwards on the reel, without rewinding after recording or playback. Fast speeds (forward or rewind) on most tape machines put uneven tension on the tape, and stronger tension than the smooth layering of play-speed tape motion. This extra strain tends to stretch and warp the tape, which over a long period of time becomes permanent. The extra tension also presses layers of tape closer together, resulting in "Print-through" of the audio on one layer to the next.

System wiring usually gets done in the haste of setting up "my new hi-fi," and the owner ends up cursing the store for poor performance when it's really his own fault. Cheap stereo will always sound like cheap stereo; but good equipment can sound cheap if it isn't hooked up properly.

Last things first: The speaker. The speaker wire you may have gotten with the system may not be thick enough. Check the spool for the gauge number of the wire—the higher the number, the *thinner* the wire. Most stores sell #22 or #24 wire as "speaker wire." This is fine if your speakers aren't more than four feet from the amp, but if they're farther away a good deal of the amplifier output will be lost in heating the wire before it even reaches the speaker. If you're planning to put your speaker six to ten feet away, #18 or #20 wire is the smallest you should use. Incidentally, that's ten feet of *wire*, not ten feet of distance. If you have to run the wire under the rug and over the sofa and three times around Aunt Martha, take all that distance into account when selecting the wire gauge.

Your local hardware store may not have "speakerwire" in #18 gauge. No sweat—ask him for some "zip cord." That's what they call lamp cord, which is usually #18 gauge or thicker, and which works fine.

The difference between "zip cord" and "speaker wire" is simply that "speaker wire" is color-coded to indicate polarity, and "zip cord" is not. Or is it?

Yes, it is. You'll find that one of the two wires in a piece of zip cord has a ridge along the outside of the insulation, or is rough instead of smooth, or has printing while the other does not. If none of these appears, there still may be a coding inside the insulation; most manufacturers include a colored thread with one of the bundles of copper strands.

Why is polarity important? When you hook up the speakers to the amp, it's possible to get them "out of phase," so that one speaker cone moves inwards as the other moves out. When that happens, the air is tossed back and forth between the two speakers, rather than pumped in and out. The effect sounds like a lack of bass, since this is most usually the part of the music fed in-phase to the speakers.

To insure an in-phase speaker connection, hook one speaker up any way you care, and make sure that you hook the other up in the same way: if the ridged/printed/threaded side of the first speaker's wire goes to the common or ground terminal on the back of the amp (which is marked "common" or "-" or "ground" or has a black connector), make sure the other is hooked up that way, too. The terminals on the back of the speaker are usually marked with plus and minus, too, and these connections should also match. If the speakers have no polarity marking, just make sure that the side of the wire going to the left-hand terminal on one speaker matches the left-hand connection on the other.

It's always possible that someone at the factory made a mistake in wiring the speakers (or the amp). If you're not certain that your speakers are properly phased, stand about four feet in front of them at dead center between them, and listen to a mono program or record. If the bass seems weak, reverse the leads of *one speaker only* at the speaker end (make sure you turn the amp off before disconnecting the wires). The connection which results in stronger bass is the proper connection.

Speaker wiring is *unshielded*—there is no metal covering over the wires to keep out stray hum and noise signals. Since speaker signals are at high level and low impedance, shielding is unnecessary. Cables between components of the system, however, are at a lower level and high impedance, which makes them sensitive to signals floating around in the air. To protect them from picking up these unwanted signals, they must be *shielded*.

A standard shielded cable consists of a wire with insulation wrapped around it, and a second layer of wire mesh wrapped around that (*illus.*). The outside wire mesh catches the outside signals and harmlessly drains them off to ground before they can interfere with the music carried on the inside wire. The outside wire mesh, called the *shield*, also interconnects the grounds of all the components of a system so that all operate at the same voltage—otherwise, you'd hear no music and a lot of hum.

At either end of a shielded cable, you'll find an RCA *pin plug*, or *phono* plug, with a center post and four fingers around it. The center post is connected to the center wire; the fingers are connected to the shield. A phono *jack* on the component accepts this plug and connects both center post and fingers to the proper part of the circuit, fingers to the ground system and center to the signal-carrying portion of the circuit.

You may still get some hum even with shielded cables, for a couple of reasons. The simplest might be that they're cheap cables, poorly shielded. The only cure for a cheap cable is to replace it. There may be a slight amount of corrosion, a film of oxidized metal on the plug or jack that insulates the plug from proper connection. This is easily cured by rubbing the plug and jack vigorously with a pencil eraser, which is abrasive enough to clean off the corrosion—but make sure the eraser crumbs are brushed out of the way before you plug it back in.

The most hasselsome problem is a *ground loop*. Take a look at the diagram. If two cables are connected between the same pieces of equipment, they form a loop around which a hum signal can flow. This loop acts just like an antenna, "broadcasting" the hum into the supposedly-shielded wires inside the cables. This is another reason that stereo cables are often made to lie together—the smaller the loop, the more chance that the picked-up hum will flow in the same direction in both shields and cancel itself out.

With a single pair of cables this usually isn't a serious problem. With a complete system (four cables to the tape deck, four to the equalizer, two from the turntable, plus a ground wire from the turntable, two between tuner and amp, plus the ground interconnection through the three-prong wall plugs), it can be murder.

If you find you do have ground-loop troubles, try bundling all the cables together, so that the picked-up hum tends to cancel itself out. This may or may not help, depending on how much farther one cable set has to travel than the other.

The ultimate solution is to break the loop. As long as there is *one* ground interconnection between components, the shields will work fine—you can cut the other shields to break the ground loop without killing the sound. Don't try to cut the shield on a regular cable, you're too likely to cut through the center wire as well. Instead, visit your local radio store and get some plain shielded wire and phono plugs and make a cable of your own (or get the kid with the thick glasses down the street to do it for you). The center wire goes to the center post of the plugs on both ends; connect the shield to the fingers *at one end only*. Leave the shield disconnected at the other end—in fact, wrap it with a layer of tape to make sure it doesn't accidentally short to anything.

If the components have three-pronged plugs, you may not need a "fully" shielded cable (shield connected at both ends) anywhere in the system—the ground through the wall plug serves nicely as the common connection. Chances are you'll need at least one "fully" shielded cable between each component (tuner to amp, amp to deck, deck to equaliser, or what have you), with the rest of the cables being "broken-shielded," disconnected shields at one end of the cable.

For best noise performance, plug the "broken-shield" cables in with the connected-shield ends at the *inputs* of each device—for example, the cable between amp and deck goes from "Rec out" on the amp (broken-shield end) to the "line input" on the deck (connected-shield end).

The signal from the turntable is *extremely* low-level, and hence very sensitive to hum pickup. There is usually a third cable from the turntable, a single ground wire that the book says to hook up to a ground post or screw on the back of the amp. It's usually a good idea to hook this wire up, but not always—try it both connected and left disconnected and use the arrangment that gives least hum and noise.

Your mother used to yell at you to clean up your room. Maybe she still does (mine does). *I* don't care where you throw your dirty clothes as long as you don't leave them draped over the equipment, blocking ventilation and getting dust in the components. Keeping the equipment clean is a good idea both for the life of the equipment and for its appearance if you ever want to trade it back in.

Coffee spills are obviously verboten—Coca-Cola is even worse, since it collects dust and will short out the circuitry. Less obvious is the effect of cigarette smoke, which leaves a corrosive tar on the circuitry that shorts it out and eats it away. If you have to smoke, blow the smoke in someone's face rather than at the equipment—preferably someone much bigger than you who will then make you stop very quickly (this message brought to you by The American Cancer Society and all your friends who don't smoke).

A few quick tips on getting the maximum life and performance from your equipment:

Most modern turntables and changers have a cuing lever, which allows you to drop the stylus gently on the surface of the record. If it has a fluid damper—which is to say, if it drops the tone arm slowly and smoothly—*use* it. If you're an out-of-work disc jockey who's been hand-dropping the needle for years and knows how to do it properly—then *definitely* use the cuing lever! I've seen you guys at work before!

Don't play a record more often than once an hour—as the stylus passes through the groove it slightly deforms the vinyl. The groove "remembers" its proper shape, and will slowly recover over a period of an hour or so. If you don't give it that hour to recover, it will eventually permanently deform and give a raucous, distorted sound to the music.

A small magnetic charge will build up on the heads and tape guides of a deck over hours and hours of playing and recording. The demagnetisers sold in hi-fi and radio stores are usually too small to do much good at removing this buildup—chances are they will leave more charge than they took away. If you're doing a *lot* of recording (ten hours or more every week), it's worth getting a professional demagnetiser—call a local recording studio and ask them where they bought theirs. Otherwise you can usually get away with ignoring it for a couple of months or so. Best idea is to get together with a couple of friends and buy one to circulate around—professional-strength demagnetisers cost from $25 to $50, but they're worth it.

If your speakers sound distorted and you know all the equipment is working properly, you're running at the limit of the amp's capability. Chances are good that the output stage is heating up, and the life of the amp may be reduced by as much as half. Turn down the volume (or the bass, or both)—you'll be surprised how easy it is to get used to lower volumes. Or, get a bigger amp.

Turn down the volume control before switching between different sources (deck/tuner/phono) or when turning the system on and off. The clicks and pops caused by switching transients can damage both speakers and amp faster than you think.

When recording a tape off record or from the tuner, adjust the record levels until the meters occasionally peak into the red region, but not so high that they are continually hitting the pin at the right-hand side. "Pinning" the meters causes the tape to saturate and distort; underdriving the tape will result in clean recording but increased noise level overall.

Don't lend your records to friends to tape—have them bring their deck over (if you don't have one) and do the taping for them yourself. Lending equipment or records is a great way to lose equipment, records—and friends.

Thanks for wading through all this with me—hi-fi is a little complicated but a lot of fun, and I hope if you've caught me striking ridiculous poses you've had the grace to ignore it, or at least to send me the negatives. You can get the most out of hi-fi—or anything, for that matter—only if you know something about the nuts and bolts of it, and it's always useful to remember that the weakest part of the system is the nut who's running the controls.

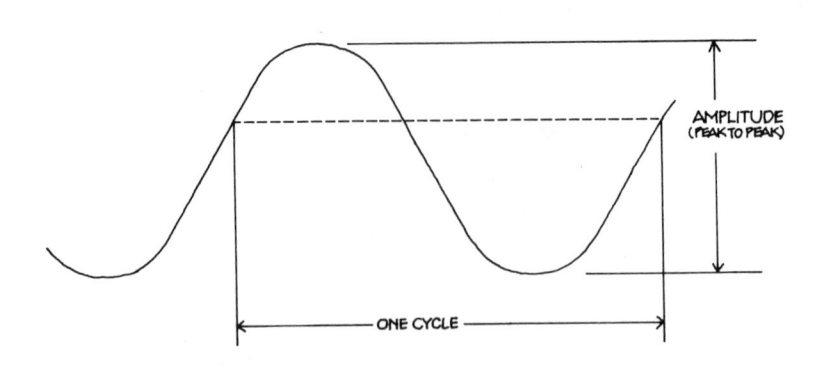

Making the Connection

AMPLITUDE
(PEAK TO PEAK)

ONE CYCLE

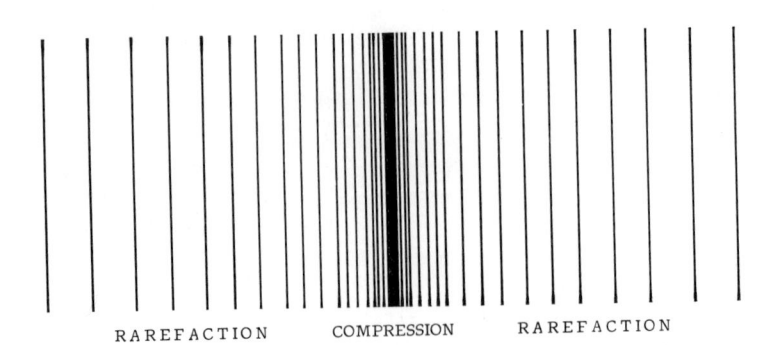

RAREFACTION   COMPRESSION   RAREFACTION

APPENDIX.

YOU DON'T HAVE TO READ THIS APPENDIX TO BUY A STEREO. It's possible to buy a good stereo intelligently without being able to recognize a decibel if you tripped over one lying in the street. Many people have.

On the other hand, knowing "why?" is always a good start towards knowing "how"—how to get the best sound that your system is capable of. It's also a good way to stay ahead of the salesman, in case he decides to try and snow you. Ergo, this appendix, your introduction to The Heavy Mysteries, and just loaded with juicy tidbits of ammunition with which to shoot down the first salesman who gets out of line.

Sound is a pattern of overpressure (*compression*) and underpressure (*rarefaction*) caused by a vibrating string, a column of air oscillating inside a trumpet tube, a speaker cone moving back and forth. Molecules in the air are alternately pressed closer to each other, and then drawn apart (diagram). These pressure waves travel through the air at approximately a thousand miles an hour—the speed of sound—and are translated into nerve impulses when they move our eardrums back and forth.

In a sense, our ears act as microphones, translating the sound energy into varying electrical nerve impulses...the higher the pressure, the higher the voltage of the impulse. Using real microphones, sound energy can be translated into varying voltages, and then processed electronically in whatever way you desire. A graph of the varying pressure (diagram) could also be a graph of the signal voltage in our electronic analogue of the sound.

These voltages can then be amplified, limited, tone-controlled, et cetera, stored (either mechanically, in the grooves of a disc; magnetically, on recording tape; or electronically, in a computer), played back, processed again, and finally turned back into sound by an *electroacoustic transducer*—which is the techies' term for a speaker.

Now that we have an electrical waveform representing our musical signal, let's push it through an amplifier and see what happens.

A French mathematician named Fourier once figured out that you can break any waveform down into component parts which are pure tones, single frequencies. A pure tone wave looks like our graph on the last page—it's called a *sinewave*, because it's also the mathematical graph of the formula y=sin x (no, it's not a coincidence, but the math gets hairier than you'd care to get into).

Our sine wave has an *amplitude*—the height of the voltage which it rises to at maximum. This is the *peak* voltage—double it for the *peak-to-peak* voltage between maximum and minimum. If you hooked this voltage across a resistor, power would be dissipated in the resistor. Imagine a battery (steady voltage) that dissipated the same amount of energy, and that battery's voltage would be the *root-mean-square* voltage. The math boys figures out that our theoretical battery's voltage must be the square root of the integral of the square of the instantaneous voltage—the root of the mean of the square. (Aren't you thrilled to know that?)

It also has a *frequency*—the number of times in a second that the voltage swings from zero to max to zero to min to zero again. This zero-to-zero variation is called one *cycle*. The higher the frequency, the higher the musical pitch of the note. Human ears can hear tones that vary from 20 cycles per second to 20,000 cycles per second.

This is a good point to take time out and learn some names. Units of scientific measurement are internationally agreed upon, so everyone will know what everyone else is talking about. The name *Hertz* was chosen to mean "cycles per second," in honor of 19th-century physicist Heinrich Hertz, who laid the foundations of modern radio theory. Thus a 400 Hz tone ("Hz" is short for "Hertz") has a frequency of 400 cycles per second.

Similarly, the man who worked out the mathematical relationship between voltage, current and resistance had named after him the unit of resistance, the *Ohm* (abbreviated with the Greek letter omega Ω . Italian experimenter Count Alessandro Volta has the unit of electrical potential, the *volt*, named after him. There are also words—prefixes, actually—for size. Scientists are often working with numbers in the millions or fractions in the millionths. Greek and Roman prefixes form a numeric shorthand, like "kilo," the prefix for "a thousand." "Mega," abbreviated "M," means a million. "Milli" is a thousandth, and "micro," (abbreviated with the Greek letter micro: μ) is one-millionth. A 2,000 Hz tone is a 2 kiloHertz, or 2 kHz tone. A phono cartridge has an output measured in thousandths of a volt, or *millivolts*.

Now we've got some vocabulary to play with. Our range of hearing extends from about 20 Hz to 20 kHz. People lose some high-frequency hearing as they grow older—the average 40-year-old may hear nothing above 12 kHz.

With nearly half his frequency response gone, you might think his hearing would be severely muffled. But because of the way frequency relates to musical pitch, it's not so—he's only lost a *tenth* of his hearing.

Here's why: Around the third century B.C., the Greek philosopher Pythagoras—the same gentleman who used to play around with squares and right triangles and hypotenusi—began plucking at musical instrument strings. He found that you could get a note one octave above the note the string was tuned to simply by cutting the string in half and letting it vibrate twice as fast. Another halving—down to quarter length this time—and you get a second octave. In thirds, the string gave a note intermediate between first and second octave.

These new notes are part of the *harmonic series*. Every time you multiply the *fundamental* frequency by a whole number, you get an *harmonic*. The elements of the harmonic series form chords based on the fundamental note. The second multiple of the fundamental is one octave above; the third, an octave and a fifth; the fourth, another octave; the fifth multiple is two octaves and a major third.

So let's figure out how many octaves we have in the 20 to 20,000 Hz range. The first octave runs from 20 Hz to its octave, 40 Hz. The second is 40 to 80 Hz; the third, 80 to 160; 160-320; 320 to 640; 640-1280; 1280-2560; 2560-5120; 5120-10240; 10240 to 20480, for a total of ten octaves. This is what we meant about only losing a tenth of his hearing—the man can hear all but the very highest octave, where very little musical energy shows up anyway.

This, incidentally, explains why white noise sounds high-pitched. *White noise* is the technical term for random energy spread evenly through the frequency spectrum. That is, the same amount of energy (on the average) appears between 20 and 40 Hz as does between 18,000 and 18,020 Hz. Thus half of all the noise energy appears in the top octave—three-quarters in the two highest octaves. Eliminate noise in this region (as the Dolby B noise reduction circuit does), and you've practically got the game licked.

Volume—intensity of sound—affects the ear similarly, in a

multiplicative rather than an additive way. Double the power in a sound and you increase the volume by a given amount. To get that same increase again, you must again double the power.

It's impractical to talk about equal increases in intensity when one is a tenth-of-a-watt increase and the other a ten-watt change. Bell Telephone engineers in the early 1900's solved this problem of comparing levels by inventing the *decibel.*

The powers-of-two series we used is an *exponential* function; to transform it to an additive (*linear*) function we have to use *logarithms.* A logarithm is an exponent that raises a base number to the desired number (huh?). Try that again: 100 is ten squared, ten to the second power. Thus the logarithm to the base 10 of 100 is 2. $1000 = 10^3$, so $\log_{10} 1000 = 3$. For numbers in between, fractional exponents are used—the log of 5 is about .6.

The Bell engineers first used the Bel (named for Alexander Graham Bell, the telephone's inventor. Don't ask me what happened to the second "l"...), defined as the log of the ratio of two power levels:

$$B = \log \frac{P_1}{P_2}$$

They soon found the size of the Bel inconvenient, and divided it into units ten times smaller—a tenth of a Bel, or a deci-Bel. Decibel.

The decibel, defined by the formula

$$db = 10 \ \log \frac{P_1}{P_2}$$

turned out very neatly to be almost exactly the smallest change of level that the human ear could detect—and therefore a nice unit to work with.

Notice that decibels always talk about power *ratios*—comparing one level against the other. Frequency response measures decibels above and below some reference level. An amp with a gain of ten decibels will make any signal ten decibels louder than it was before. But you can't say "How many decibels does that amp have?" like you can say "How many watts?" Decibels must always refer to some other reference level.

We've seen that power increases exponentially as decibels go up linearly. That explains the modern trend toward superpower amplifiers. Suppose you have a twenty-watt amp driving your speakers and you're not happy with the level you're getting. Adding three decibels means doubling your power—and three decibels isn't much increase in volume. Three more decibels? Double it again—already we're up to eighty watts, and we've only gained six decibels. To get realistic rock concert volumes from most speakers (if they'll stand the strain) you need upwards of 150 watts per channel... which may help explain why improving your stereo gets harder and more expensive as you go higher in quality.

GLOSSARY.

A

Acoustic suspension—A technique for getting good bass out of relatively small speaker boxes. The speaker cone is very loosely suspended, and the springiness of the trapped air inside the sealed box is depended on to return the speaker cone to center after a signal is passed through the speaker.

Amplitude modulation—Process whereby a radio signal's strength is varied in proportion to a modulating signal for transmission of that signal.

Antenna—A device for collecting a radio signal from the air and sending it to the receiver.

Anti-skate—A device to counteract the tendency of a tone arm to move to the center of the record as the record spins. See *"skating force."*

B

Bandwidth—The number of Hertz over which a signal appears, or over which a device operates. Human hearing has a bandwidth of approximately 20,000 Hertz.

Bass—Low frequencies, usually considered to be from 20 hz. to about 300 hz.

Bass reflex—Open-box or ported-box speaker design in which sound from the back of the speaker cone is channeled around and out to reinforce sound from the front of the speaker cone.

Biamplification—Separate amplification of low and high frequency components of a single signal before feeding to separate speaker elements. More expensive but more efficient than passive crossover elements. See *"crossover."*

Breathing—Momentary rise in background noise during quiet passages of a compressed signal. See *"compression."*

C

CD-4—Quadrophonic recording technique developed by Japan Victor Corporation. Uses sum-and-difference techniques and frequency-modulated supersonic carrier to put four-channel signal with good separation into a two-channel record groove.

Capture effect—A property of frequency modulated signals whereby an FM receiver only demodulates the stronger of two signals on the same frequency.

Capture ratio—The required difference in signal strength for the capture effect to take place.

Ceramic—A type of phono cartridge using synthetic material which generates a voltage when squeezed by the needle's motion. See *"crystal."*

Clipping—Distortion caused by overdriving an amplifier. The output rises to the maximum possible level and stays there until the signal drops to a range within the capabilities of the amplifier, thereby "clipping" the peaks off the amplified waveform.

Compatible—Standards which work acceptably in two or more modes. Stereo FM transmissions are compatible with mono reception, for example, because no information is lost by the mono listener.

Compression—Reduction of the dynamic range of a signal by means of an automatic-gain-control amplifier. Radio stations often use compression to insure maximum modulation of their signals.

Conical—Phono stylus shape whose cross-section is circular.

Crossover—Device for separating frequency ranges of a signal and sending them to the proper speaker elements for reproduction.

Current—The flow of electrons through a conductor comprising an electrical signal. See *"voltage."*

Crystal—Type of phono cartridge using a property of some crystalline materials which generate a voltage when placed under pressure.

D

db.—Decibel.

Decibel—A unit for measuring the ratio between two power (or voltage or current) levels. Roughly the smallest change in volume that the human ear can detect.

De-emphasis—High-frequency reduction in FM broadcast reception used to reduce the effects of noise. See *"pre-emphasis."*

Distortion—A change in the waveform of an amplifier's output from the form of the input signal. See *"harmonic distortion," "intermodulation distortion."*

Dolby—Trademark for a noise-reduction circuit used in tape recording, developed by Dr. Ray Dolby.

Doubling—The tendency of an over-driven speaker to vibrate at twice the frequency being fed to it.

Driver—The sound-generating element in a horn speaker.

Dynamic—Speaker which uses magnetic charges to move the cone.

Dynamic range—Decibel difference between loudest and softest passages of a signal.

E

Electrostatic—A type of speaker which uses attraction and repulsion between electrically-charged membranes to generate sound. See *"dynamic."*

Elliptical—Stylus shape whose horizontal cross-section is an oval or ellipse.

Expander—Gain-controlled amplifier which restores or increases dynamic range of a signal, usually after compression.

F

Fair Trade—Laws in some states regulating the minimum retail price which can be charged for an item. Intended to prevent large retailers from putting smaller competitors out of business through price-cutting, they have largely backfired by eliminating competition that would reduce prices to the customer, and are being considered now for repeal.

Filter—Circuit which eliminates an unwanted signal component, such as high frequencies (scratch filter), lows (rumble filter).

Flutter—Rapid variation in the speed of a turntable or tape deck, causing the pitch of playback to vary up and down with a tremolo-type effect. Most noticeable on guitars, piano, acoustic instruments.

Frequency—Measure of the number of cyclic variations a waveform goes through in a second. The higher the frequency of a signal, the higher its musical pitch.

Frequency modulation—Process whereby a radio signal's frequency is varied in proportion to a modulating signal for transmission of that signal.

Frequency response—Measure of an amplifier's performance over the audible frequency range. Measured as constant gain (within specified decible limits) over given bandwidth. Example, $\pm$ 1 db, 20 to 20,000 hz.

Full-logic—Matrix quad decoder with gain-adjusting elements for apparent improvement of channel separation.

Fundamental—Lowest (root) frequency of a harmonic series. If you have a fundamental note of 440 hz (concert A), you will have harmonics at 880 hz., 1320 hz., etc.

G

Gain—Increase in voltage or power from input to output of an amplifier. Measured in decibels.

Graphic equalization—Adjustment of individual frequency bands (instead of simple bass and treble) across audible range to compensate for deficiencies in room acoustics, program material, etc.

H

Harmonic—Signal at an integral multiple of a fundamental frequency. In music, octaves and major triads are related harmonically to the root note.

Harmonic distortion—Distortion which generates harmonics of the input frequency.
Heterodyne—To mix two frequencies in an amplifier designed to produce intermodulation distortion, so as to obtain a new signal at the sum or difference of the original frequencies.
Horn—Type of speaker using a long flaring tube in front of the cone to improve efficiency and reduce distortion.
House brand—Component made for a large retailer and usually listed at an inflated price to create the illusion of huge discounts in a system purchase.
Hum—Low-pitched signal or buzz caused by pickup or improper filtering of line voltage from the wall plug.
Hz—Abbreviation of Hertz. Unit of frequency; cycles per second.

I

I.H.F.—The Institute of High Fidelity, a manufacturer's group which has standardized a set of specifications.
Impedance—Measure of the degree to which a component or device allows the flow of an alternating current, such as a musical signal.
Infinite baffle—Closed-box speaker which traps sound off the back of the speaker cone inside the box. Acoustic suspension speakers are one type of infinite baffle.
Integrated amplifier—Component which includes preamplifier and power amplifier in one cabinet.
Integrated circuit—Electronic component including a complete circuit within a single "can" or package, for reduced size and weight.
Intermediate frequency—The frequency to which radio signals in a tuner or receiver are heterodyned for processing and demodulation.

K

Kilo—Prefix indicating "a thousand." Abbreviated *"k."*

L

Logarithm—A means of expressing a number as a power of a "base" number. For example, ten raised to the second power ($10^2 = 100$) is one hundred, so the logarithm (to the base ten) of one hundred is two.
Logic—Electronic devices which use on or off states to indicate numbers or conditions. See *"full-logic."*
Loss leader—A product priced below cost or below its usual price so as to attract customers to the store.
Loudness—Increase in bass and treble at low volumes to compensate for the human ear's reduced sensitivity to bass and treble at low levels. Sometimes used incorrectly to indicate volume. See *"volume."*

M

Magnetic—Type of phono cartridge which uses moving wire coils or magnets to generate a signal proportional to the motion of the stylus.
Matrix—Technique for reducing four channels to two without increasing signal bandwidth, but with reduced channel separation after decoding.
Mega—Prefix indicating "one million" abbreviated "M."
Micro—Prefix indicating "one one-millionth, abbreviated with Greek letter micro, " μ ."
Midrange—Frequencies between 300 and 5000 hz., approximately. Also, a speaker element designed to reproduce these frequencies.
Milli—Prefix indicating "one one-thousandth." Abbreviated *"m."*
Mix—To combine two or more signals into a single signal, especially in recording. Also, the signal resulting from this process. In receiving, to heterodyne two signals together to produce a third intermediate frequency.
Mode—Mono, stereo, or quadrophonic.
Modulation—Modifying some characteristic of a radio signal so as to code that signal for decoding at the receiving end. Also, the message so coded.
Monitor—To listen to a signal for quality as it is being recorded. Also, any device for that listening.
Monophonic—A single musical signal complete and entire in itself.
Multipath—Interference caused by reception of a single radio signal direct from the transmitting antenna and simultaneously from a reflected path (off a mountain, airplane, or building, for example).
Multiplex—Technique for coding two or more signals into a single channel

N

Noise—Electrical energy generated by the random thermal motion of molecules in electronic components. *See also "white noise" and "pink noise."* Note: tape hiss, which sounds similar, is caused by random variations in magnetic charge from particle to particle of the tape oxide.

Noise reduction—Circuitry designed to eliminate or reduce noise or tape hiss in an audio signal. It can be post-processing (cleaning up an already dirty signal) or complementary (modifying a signal before recording and then remodifying after recording back to the original signal, reducing the noise effects of the recording process. Dolby equipment is complementary.)

O

Ohm—The unit of resistance, named after the German physicist who discovered the mathematical relationship between resistance, current and voltage.

Overdub—In multi-track recording, to add a new track after the basic tracks have been laid down.

Oxide—The magnetically-active layer of recording tape, on which the signal is recorded. Various oxide compounds of iron and chromium are used.

P

Peak—The maximum instantaneous value of current, voltage or power for a varying waveform.

Peak-to-peak—The range of voltage or current from most positive to most negative of a varying waveform.

Phase—The fraction (expressed in degrees out of 360°) of a cycle through which a wave has passed. If two signals of equal strength and frequency are 180° out of phase with each other (that is, one is going positive when the other is going negative), they will cancel each other out when combined. Speakers which are connected "out of phase"—that is, with differing polarity—will tend to cancel out the bass, leaving a weak and trebly sound. Phase is also important in matrix quad processing and in multiplex stereo reception.

Phone plug—A type of audio connector consisting of a barrel of metal approximately 1½ inches long and ¼" in diameter. The tip of the plug is electrically isolated from the sleeve; in stereo connections, a third isolated *ring* may also be present. Each piece (tip, ring, sleeve) makes a separate connection electrically.

Phono plug—A smaller audio connector consisting of a small pin approximately ½" long and 1/8" in diameter, surrounded by metal fingers cylindrically placed, approximately ½" in diameter. Developed by RCA in the 1950's, its proper name is an *RCA pin plug*. (most people get phone and phono plugs confused. Even some engineers I know...don't let it get you down.)

Phono preamp—High-gain amplifier with specially-tailored frequency response for amplifying low-level signals from a magnetic-reluctance cartridge to a level where they can be processed by the control-preamplifier circuitry. *See also "RIAA."*

Piezoelectric—A property of some crystals and ceramic materials whereby a voltage is generated when the crystal structure is flexed. Used in inexpensive phonographs.

Pilot—The 19 kHz signal transmitted on a stereo FM broadcast that synchronizes the stereo demodulator for proper decoding in an FM receiver. Also, a panel lamp showing that equipment is operating.

Pink noise—Random noise whose frequency characteristic is such that each musical octave contains the same amount of energy on the average as each of the other octaves. *See "white noise."*

Platter—The rotating part of a turntable.

Power amplifier—The section of a system which provides high-current drive to the speakers.

Preamplifier—The volume/tone-control section of a system, usually incorporating the input selector, phono preamplifier and other controls. Preamp and power sections may be separate, or combined in a single component called an "integrated amplifier." If tuner, preamp and power sections are contained in a single component it is called a "receiver."

Pre-emphasis—The treble boost applied to an audio signal before FM broadcast transmission. A corresponding *de-emphasis* of the treble is performed in the FM receiver. This processing tends to reduce the audible effect of noise picked up in reception of the FM signal (*see "noise reduction."* This is a simple complementary n.r. system).

Psychoacoustics—The brain doesn't always hear what the ears pick up. If two sound sources of equal pitch and intensity are placed some distance apart in front of a listener, he will "hear"—that is, his brain will interpret the sound as—a single broad source located somewhere between the two actual sources. Noise signals loud enough to hear will be "masked" by louder signals in the same frequency range. These and other effects are studied by designers of hi-fi equipment; their study is the science of psychoacoustics.

Pumping—Variations in sound level caused by insufficient power in the power supply of an amplifier.

Q

Quad—Reproduction of a sound through four separate but related audio signals for purposes of creating spatial placement effects. *See "stereo."*

R

RIAA—The equalization curve chosen as standard by the Recording Industries Association of America. In a process similar to pre- and de-emphasis in FM broadcasting, treble notes are boosted before being recorded in the grooves of a record, then cut correspondingly on the phono preamp for playback. The effect is to reduce the level of surface noise in the disc.

RMS—Root-mean-square. The RMS value of a waveform is the equivalent of the constant DC voltage (or power) which would generate the same amount of energy over the same period of time. Mathematically, it is the square root of the integral of the square of the instantaneous voltage (or power) over a given interval of time, divided by the length of that interval.

In power specifications, the RMS power is the power that the amp can supply with a continuous tone, usually for a long period (five minutes to an hour or more).

Rejection—The amount by which a given circuit filters out an unwanted signal component. Also, what happens when you go to the bank for a loan to buy your system.

Resistance—The amount of blockage presented to current flow. The voltage required to generate a given current is proportional to the resistance that the current must flow through (*Ohms law*).

Reverberation—If the ear hears a sound first directly, then reflected off another surface, it may not distinguish between the two. If the reflected sound reaches the ear less than a tenth of a second after the original, the two sounds will blend into a single unit in the brain, and the second sound is called a *reverberation*. If the sounds are more than a tenth of a second apart, the ear distinguishes them separately and the second sound is an *echo*. Recording studios use acoustic chambers and mechanical devices to generate reverberation effects. Note: when a signal is recorded on a three-head tape deck, and the playback head signal is fed back to the recording head, producing a recurring loop of sound, the process is often mistakenly called *"reverb."* The correct name for the technique is *slap echo*.

Rolloff—The frequency response of an amplifier does not cut off sharply at a given "top" or "bottom" frequency, but gradually decays around that frequency. This gradual change of gain is called *rolloff*. (Certain equalization and tone-control circuits may be designed to roll off more quickly than others.)

Room acoustics—the collective effect of room size, materials, furnishings and environment on the characteristics of sound in the room. The pattern of reflection and absorption of sound can drastically change frequency response, stereo placement effects, etc. for the listener.

Rumble—The low frequency noise caused by irregularities in the platter and drive system of a turntable.

S

SCA—Subsidiary Communications Authorization. In addition to the normal stereo signal, FM broadcast stations may be licensed to carry a third signal which is rejected by home receivers but can be picked up by special equipment. This channel is usually used for background music, although some stations are now using it for pocket pagers, remote metering of their distant transmitters, and other purposes.

Saturation—A given particle of oxide on recording tape can only retain a certain maximum magnetic charge. When the recorder attempts to put more signal on the tape than that particle can hold, the tape is *saturated*. The audible effect of tape saturation is similar to clipping in an amplifier.

Scratch filter—Circuit designed to roll off high frequencies of an audio signal to eliminate treble noise from tape or surface noise on a disc.

Sel-Sync—Ampex trademark for a system where selected tracks of a multi-channel recording can be monitored from the recording head rather than the playback head. This allows subsequent recording in synchronization with the original music. Though it is a trademark, and may be called by other names by other manufacturers, "sel-sync" is generally used in the recording industry to indicate this technique.

Selectivity—Measure of a tuner's ability to reject signals near to the desired frequency.

Sensitivity—Measure of a tuner's ability to receive weak signals. Sometimes used to denote the input voltage range for a high-gain preamplifier (microphone or phono).

Separation—The amount of leakage of one channel's signal into another channel in stereo or multi-track recording.

Shibata—Stylus developed by JVC for playback of CD-4 recordings and named for the leader of the JVC development team. A Shibata stylus has a thinner horizontal cross-section than an elliptical stylus for superior tracking, but compensates for the smaller groove contact area with a flatter vertical cross-section.

Shielding—Metal or metal-mesh surrounding signal-carrying circuitry, which blocks outside signals and noise from interfering.

Sibilance—Harsh whistling sound caused by distortion of high-frequency sounds such as voiced *"s."*

Signal-to-noise ratio—Relative level of inherent circuit (or device) noise to normal high-level audio signal, expressed in decibels. For broadcast signals it is usually measured relative to full modulation level; for recording, usually relative to a specified level below clipping or saturation.

Sine wave—The waveform of a single-frequency signal. Mathematically, it is the graph of the function $y = \sin x$, or the height of a given point on an evenly-rotating wheel graphed against time.

Skating force—When the groove on a disc passes under the stylus at an angle to the axis of the cartridge, the stylus tends to be pushed towards the center of the record. This slight pressure inward is the *skating force*.

Slap-echo—Feeding the signal from the playback head of a three-head tape deck back to the recording head, thus creating a "loop" of sound that slowly decays, is called *slap echo*. See *"Reverberation."*

Squawker—Seldom-used term for the midrange speaker element in a three-way speaker system.

Spurious response—Unwanted signal component that is not completely filtered out. In properly-designed equipment, spurious responses can be detected with test equipment but are well below audible level.

Stereo—Two-channel reproduction of a single sound or performance for spatial effects.

Stylus—The diamond tip of a phonograph cartridge which rests in the disc groove. Generally used to denote the entire structure which holds stylus in place in the cartridge body.

Subcarrier—Single-frequency signal which is modulated to carry stereo or SCA information in FM broadcasting. The modulated signal, along with the main channel signal and any other subcarrier signals, is then used to modulate the transmitted carrier frequency at the station's assigned broadcast frequency.

Suppression—Rejection of an unwanted signal component.

Synthesizer—Device which creates the electrical analogue of sound waves directly, rather than transducing them from real sound waves.

T

Tails (or tails out)—A tape which is left on the takeup reel after recording or playback, with the end of the tape at the outer edge of the reel. The tape is subjected to more even tension in this manner than by fast-forwarding or rewinding.

Tanstaafl—"There's no such thing as a free lunch." See *"trade-off."*

Tape monitor—Switching arrangement in a control amplifier which allows monitoring from the output of a tape deck while recording from another source.

Three-way—A speaker system which uses separate speaker elements to reproduce the bass, midrange and treble ranges.

Tone—The frequency characteristics of an audio signal; or, controls used to adjust these characteristics; or, a single-frequency signal used for test purposes in audio equipment.

Tone arm—The device which holds the phono cartridge parallel to the record surface, with the stylus touching the groove walls.

Track—In tape recording, the fractional width of the tape used to record one channel; in broadcasting, one selection on a record (also called a *"cut"*).

Trade-off—An engineering concept which states that improvements in one characteristic of a circuit are accompanied by deterioration of another characteristic. For example, improvements in the overload capability of a transistor stage are usually accompanied by increases in the inherent noise of the stage. A very familiar trade-off is performance vs. cost.

Transducer—A device for transferring energy from one type to another. A microphone or phono cartridge is a mechano-electrical transducer; a speaker is an electro-mechanical transducer.

Transient—A sharp impulse or change in energy level, such as a pop or a click. Because of their sudden change, transients put the greatest demands on hi-fi systems, or any physical system.

Treble—The high-frequency range of sound, usually considered to be above 3 kHz or so.

Turntable—A device for spinning a recorded disc and playing back the signal on it.

Tweeter—A speaker element designed to reproduce the treble range.

Twinlead—A flat, ribbonlike two-conductor cable used for connecting TV or FM receivers to an outside antenna.

Two-way—A speaker system which uses only two elements (or groups of elements) to reproduce the entire audio range, bass-lower-midrange, and upper-midrange-treble.

V

Voice coil—The coil of wire through which the power amplifier output current is passed. This current sets up a varying magnetic field which, interacting with the field from the speaker magnet, moves the attached speaker cone, generating sound.

Voltage—Electrical "pressure"; the strength of the force which causes electrons to flow through a circuit.

Volume—The acoustic level of an audio signal. *See "loudness."*

W

Waveform—The shape of a graph of voltage as it varies with time in an electrical signal. The simplest periodic waveform—a sine wave—corresponds to the simplest audio signal, a single-frequency tone.

Whistle filter—Circuit built into some AM tuners and receivers which rejects the 10 kHz beat note between two AM stations on adjacent frequencies.

White noise—Random electrical energy whose frequency distribution is such that average energy in a given bandwidth is the same at all frequencies (for example, the energy between 20 and 40 Hz is the same as the energy between 16580 and 16600 Hz). *See also "pink noise."*

Wire gauge—A standard for measuring the diameter of electrical wire. The higher the gauge number, the thinner the wire.

Woofer—A speaker element designed for reproducing the bass range.

Wow—Slow cyclic variation in the speed of a turntable or tape deck. It causes a siren-like effect on the pitch of played-back signals.